SpringerBriefs in Service Science

SpringerBriefs present concise summaries of cutting-edge research and practical applications across a wide spectrum of fields. Featuring compact volumes of 50 to 125 pages, the series covers a range of content from professional to academic.

Typical publications can be:

A timely report of state-of-the art methods

A bridge between new research results, as published in journal articles

A snapshot of a hot or emerging topic

An in-depth case study

A presentation of core concepts that students must understand in order to make independent contributions

SpringerBriefs are characterized by fast, global electronic dissemination, standard publishing contracts, standardized manuscript preparation and formatting guidelines, and expedited production schedules.

The rapidly growing fields of Big Data, AI and Machine Learning, together with emerging analytic theories and technologies, have allowed us to gain comprehensive insights into both social and transactional interactions in service value co-creation processes. The series SpringerBriefs in Service Science is devoted to publications that offer new perspectives on service research by showcasing service transformations across various sectors of the digital economy. The research findings presented will help service organizations address their service challenges in tomorrow's service-oriented economy.

Hui Yang • Bing Yao

Sensing, Modeling and Optimization of Cardiac Systems

A New Generation of Digital Twin for Heart Health Informatics

Springer

Hui Yang
Industrial and Manufacturing Engineering
The Pennsylvania State University
University Park, PA, USA

Bing Yao
Industrial and Systems Engineering
The University of Tennessee Knoxville
Knoxville, TN, USA

ISSN 2731-3743 ISSN 2731-3751 (electronic)
SpringerBriefs in Service Science
ISBN 978-3-031-35951-4 ISBN 978-3-031-35952-1 (eBook)
https://doi.org/10.1007/978-3-031-35952-1

This Springer imprint is published by the registered company Springer Nature Switzerland AG
The registered company address is: Gewerbestrasse 11, 6330 Cham, Switzerland

Preface

Heart disease is one of the leading causes of death in the world. Heart health is indispensable to improving the wellbeing and economics of our society. Therefore, healthcare systems are investing heavily in digital health technologies to increase information visibility and cope with the disease complexity in the United States and around the world. Indeed, advanced and pervasive sensing bring the new generation of internet of health things (IoHT), which provides smart and interconnected monitoring and control of physiological dynamics. As such, massive data become readily available in the healthcare environment. This provides an unprecedented opportunity to build new cyber-physical systems for the heart, or called "cyber-heart systems," "digital twin of the heart," "Internet of Hearts."

There may be different names or terminologies for the integration of analytics in the cyber world with human heart in the physical world. The fundamental concept stays the same as follows: Cardiac dynamics of human subjects are reflected in the cyber space through advanced sensing, data-driven information processing, and computer modeling. Analytics in the cyber space (e.g., artificial intelligence and machine learning) exploits the acquired knowledge and useful information from data for the feedback loop of optimal actions (or healthcare interventions) to the physical world, e.g., inferring disease pathology from sensor observations for personalized care delivery. This movement toward cyber-heart digital twin holds great promise to improve heart health through sensor-based modeling of cardiac dynamics, data-enabled health innovations, and personalized optimization of health management.

This book provides an overview of our continuous research efforts on the proposed **"sensing-modeling-optimization"** framework to build new cyber-heart digital twin for improving the delivery of healthcare services. Our research is aspired by long-endured "pains" of clinical collaborators in handling rich data streams, as well as their "needs" to exploit data-driven knowledge and simulation models for optimizing healthcare interventions. Our ultimate goal is to usher in the new cyber-heart digital twin technology and contribute to improving healthcare outcomes, saving human lives, as well as reducing healthcare expenses through optimal medical treatment and planning. We hope this book will serve as a reference for healthcare researchers, practitioners, and students and further help catalyze more in-depth investigations and multi-disciplinary research efforts to bring a heart-healthy future and make a lasting impact on our society.

This book is organized into three research topics: (i) "**Sensing**" covers the state-of-the-art in ECG signal processing for the identification of disease-altered cardiac dynamics. (ii) "**Modeling**" presents personalized cardiac modeling through integrating the physics-based knowledge of the cardiovascular system and machine learning of multi-source medical data. (iii) "**Optimization**" introduces the integration of simulation models with sensor-based data fusion for optimizing the process of surgical planning. The simulation optimization approach provides a unique opportunity to search for optimal actions with the "virtual" heart, as opposed to traditional "experience-based," "trial-and-error," or subjective decisions in the real-world heart. The "sensing-modeling-optimization" approach is significant because it is a major step in a continuum of research that is expected to address key questions of disease progression (e.g., medical questions, pharmaceutical design questions, or surgical questions).

In addition, this book includes extensible modeling software packages that can be useful not only to the optimization of medical decision-making, but also to investigate the vast array of issues surrounding heart diseases and their treatments. The URL links for software packages (also available via the Matlab FileExchange) are listed as follows:

- Toolbox of EKG Animation and State State Representation: https://www.mathworks.com/matlabcentral/fileexchange/58238-electrocardiogram-animation-and-state-state-representation
- Toolbox of EKG ensembles and waterfall visualization: https://www.mathworks.com/matlabcentral/fileexchange/77993-ensembles-and-waterfall-visualization-of-ecg-heart-beats
- Toolbox of VCG animation with the colored speed: https://www.mathworks.com/matlabcentral/fileexchange/58241-vectorcardiogram-vcg-animation-color-speed
- Toolbox of spatiotemporal dynamics simulation in the whole heart: https://www.mathworks.com/matlabcentral/fileexchange/78194-simulating-spatiotemporal-dynamics-in-the-whole-heart
- Toolbox of recurrence plot and recurrence quantification analysis: https://www.mathworks.com/matlabcentral/fileexchange/58246-tool-box-of-recurrence-plot-and-recurrence-quantification-analysis

We are putting together continuous efforts to expand and share software packages via the public repositories and personal websites. We are also welcoming researchers and scientists around the world to leverage the resulting software to study issues on a variety of heart diseases and treatments. Any comments or suggestions to improve the dissemination of software tools are also appreciated. Finally, we are looking forward to working and collaborating with laboratories worldwide to promote the "sensing-modeling-optimization" paradigm and build the new technology of cyber-heart digital twin.

Lastly, we would like to thank all of former students, advisors, and collaborators for their willingness to work with us, mentor us, and encourage us in research and education. We also gratefully acknowledge the support of the National Science Foundation under Grants IOS-1146882, CMMI-1266331, CMMI-1454012, IIP-1447289, CMMI-1617148, CMMI-1619648, CMMI-1646660, IIP-1447289, and MCB-1856132. In addition, we acknowledge NSF Rapid grant IIP-2026875 and NSF I/UCRC Center for Healthcare Organization Transformation (CHOT) and NSF I/UCRC award #1624727 for supporting our research in the healthcare domain. Finally, the author (HY) thanks Pennsylvania State University, Fulbright Finland Foundation, and VTT Technical Research Center for financial support to this research during his sabbatical leave in the year 2022–2023. Any opinions, findings, or conclusions found in this paper are those of the authors and do not necessarily reflect the views of the sponsors. Finally, we thank the support and encouragement of Dr. Robin Qiu, series editor, and Jialin Yan, Springer editor, toward the completion of this book; and Springer staff for editorial and production assistance.

University Park, PA, USA — Hui Yang
Knoxville, TN, USA — Bing Yao
February 2023

Contents

Chapter 1
Introduction

Cardiovascular disease is a leading cause of death in the world. According to Center for Disease Control, it is reported that one person dies every 34 s from heart diseases in the Unites States. In 2020, there are approximately 697,000 people died from heart disease in the USA, i.e., approximately 1 in every 5 deaths [1]. The annual economic costs, direct and indirect, related to heart diseases are around $229 billion in the Unites States [1]. To improve cardiac care services and patients' quality of life, it is indispensable to sense, detect heart diseases early and further optimize medical decision making.

To increase information visibility and cope with the disease complexity, modern healthcare systems are investing heavily in physiological measurements, wearable sensing, and computing technology. As a result, massive data are readily available in the clinical environment. Realizing the full potential of rich data streams for optimal decision support depends to a great extent on the advancement of information processing and computational modeling methodologies. On the other hand, medical scientists conduct electrophysiological experiments to gain a better understanding of the physics in disease progression and how it is correlated with the cardiac behaviors. Undoubtedly, experiments are critical in the discovery process of system physics and knowledge innovation, also see Fig. 1.1. However, it is difficult to conduct the physical experiments in the human heart. Hence, scientists and engineers have been seeking alternative methods, e.g., physical experiments with animal subjects and computer experiments with simulation models.

Indeed, there are many practical and ethical limitations in the physical experiments of animal subjects. Also, it is very expensive and difficult to comprehensively investigate the disease progression across biological scales, i.e., from ion channels to cells to tissues to the whole heart. The research communities, including Physiology, Cardiology, and Engineering, have identified the urgent need to develop computer models of cardiac processes at more than one biological scale. Progress in predictive medicine can be greatly accelerated by coupling computer experiments with the wealth of data from physical experiments.

H. Yang, B. Yao, *Sensing, Modeling and Optimization of Cardiac Systems*,
SpringerBriefs in Service Science, https://doi.org/10.1007/978-3-031-35952-1_1

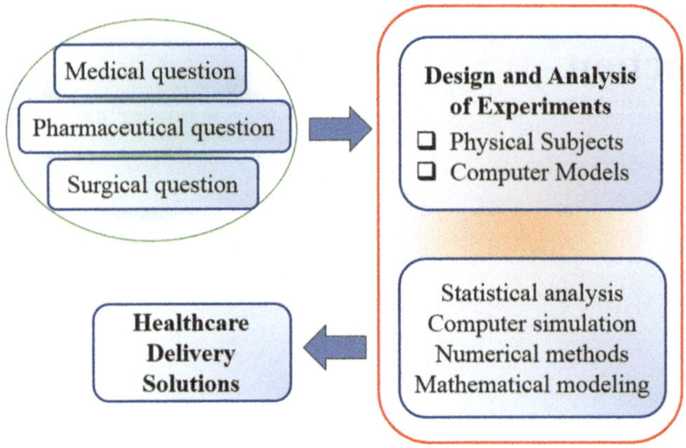

Fig. 1.1 Design and analysis of experiments for the discovery of system physics and knowledge innovation

This book will highlight recent research advances in machine learning, physics-based modeling, and simulation optimization to fully exploit medical data for promoting data-driven and simulation-guided diagnosis and treatment of heart disease. Specifically, this book will focus on three major topics:

1. **Simulation Modeling of Cardiac Systems**: This chapter will review recent advances in personalized cardiac modeling through integrating the physics-based knowledge of the cardiovascular system and machine learning of multi-source medical data.
2. **Sensor-based Modeling and Analysis of Cardiac Systems**: This chapter will review the current state-of-the-art in ECG signal processing for the identification of disease-altered cardiac dynamics.
3. **Simulation Optimization of Medical Decision Making**: This chapter will introduce the integration of simulation models with sensor-based data fusion for optimizing the process of surgical planning. The simulation optimization approach provides a unique opportunity to search the optimal medical decisions with the "virtual" heart, as opposed to traditional "experience-based," "trial-and-error," or subjective decisions in the real-world heart.

This book contributes to provide better understanding of disease-altered cardiac electrical dynamics through the sensing–modeling–optimization approach. As shown in Fig. 1.1, this is significant because it is a major step in a continuum of research that is expected to address key questions of disease progression (e.g., medical questions, pharmaceutical design questions, or surgical questions). Gaining a deeper understanding of a functional role for disease progression in electrical signaling brings broad scientific impact. For example, the early detection of disease in the initial stage can be achieved. Second, medical decisions can be optimized, and the life-saving interventions can be delivered in a timely manner. Further, "rescue" therapies can be developed by imposing molecular changes that counterbalance the dysfunction due to disease progressions. In addition, this book will include an extensible modeling software package that can be useful not only to the optimization of medical decision making, but also to investigate the vast array of issues surrounding heart diseases and their treatments.

1.1 Cardiac Electrical Signaling

A correctly beating heart is important to ensure adequate circulation of blood throughout the body. Normal heart rhythm is produced by the orchestrated conduction of electrical signals across multiple scales of cardiac processes (i.e., ion channels, cells, tissues, and the whole heart). Electrical signals in the cellular level are called action potentials (AP, see Fig. 1.2b). The action potential is produced by the concerted functions of transmembrane proteins, called ion channels (see Fig. 1.2a). Slight changes in ion channels can alter the AP waveforms and lead to changes in conduction across the heart, thereby impacting cardiac function. It is established that the types and numbers of ion channels that are responsible for producing the cardiac AP are affected by heart diseases. This is usually called arrhythmogenic reassembling (named "remodeling" in physiology) of ion channels. The reassembling process occurs with many types of heart disease [2], including myocardial infarctions, heart failure, and cardiac arrhythmias.

When a cardiac cell is stimulated, the activation and inactivation of ion channels bring the ions (e.g., Na+, Ca++, and K+) across the membrane. As shown in Fig. 1.2a, the influx and efflux of ions generate electrical currents through the ion channels. Such highly regulated ionic currents collectively produce the cardiac AP with 5 distinct phases (denoted as phase $0-4$ in Fig. 1.2b). The conduction of APs on cardiac tissues (e.g., 600×600 cells, see Fig. 1.2c) can form a spiral wave, which

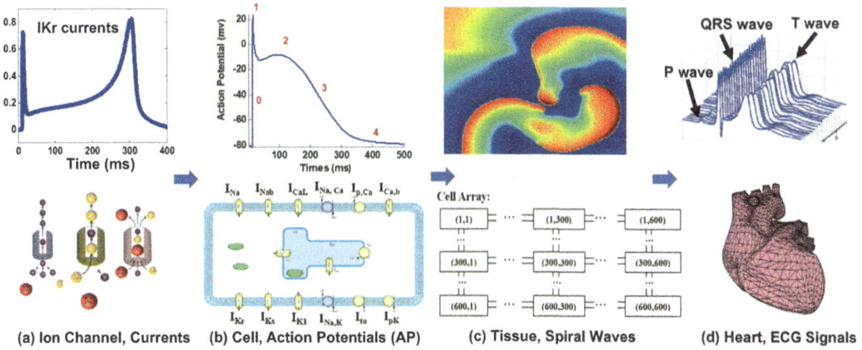

Fig. 1.2 Illustrations of multi-scale cardiac processes and conjugate electrical activities. (**a**) Ion channel and currents [3]. (**b**) Cell and action potential (AP). (**c**) Tissue and spiral waves. (**d**) Heart and ECG signals [4]

constitutes self-sustaining arrhythmias. The dynamics of spiral waves are often perturbed due to the presence of non-excitable obstacle (e.g., injured or dead tissues due to myocardial infarctions). The electrocardiogram (ECG) (see Fig. 1.2d) is the system-level dynamic behaviors that represent the integrative electrical functions of ion channels, cells, tissues, and the whole heart. There is an urgent need to investigate how the progression of diseases changes cardiac electrical dynamics across physical scales. This is conducive to perform the early detection of the onset in cardiovascular diseases and the timely delivery of healthcare solutions.

1.2 Spatiotemporal Heterogeneity of Heart Diseases

Cardiovascular diseases can take place at different locations within the heart and can progress over time. For example, the occlusion of coronary artery causes myocardial infarctions (also known as heart attacks). This occlusion leads to insufficient blood and oxygen supply that damage the cardiac tissues and trigger the cellular degradation process. Because blood vessels spread all over the heart, the anatomical locations of infarctions may include anterior, inferior, posterior, inferior-lateral, or posterior-lateral areas of the heart. The temporal triad of myocardial infarction is ischemia, injury, and necrosis. Ischemia is due to reduced blood supply, injury indicates acuteness of infarct, and necrosis is the premature death of cardiac cells and tissues. The dead tissues are usually surrounded by progressively less ischemic regions with gradual transition to healthy myocardium.

Myocardial infarctions cause substantial changes in the K+, Na+, and Ca++ currents. The arrhythmogenic reassembling of ion channels leads to important changes in cellular electrical activity and impulse propagation over days and weeks. For example, most ventricular myocytes (i.e., cardiac muscle cells) die within the necrotic zone of myocardial infarction, leaving a surviving subendocardial Purkinje fiber layer with prolonged action potentials and enhanced automaticity. Surviving myocytes in the viable border zone adjacent to a prior infarction show reduced excitability. Timing is very important to prevent life-threatening events in heart diseases. The ischemia is reversible if there are no further permanent damages to the heart muscles. However, most existing studies focus on the binary interpretation of cardiac conditions, i.e., asking whether the heart is healthy or ill. Few, if any, previous modeling methods can effectively capture the spatiotemporal characteristics of cardiac disease process and the altered cardiac electrical dynamics.

1.3 Multi-scale Modeling of Cardiac Systems

The heart is one of the most intricate organs in the human system. Computer models facilitate the quantitative simulation, elucidation, and understanding of heart functions in health and disease. In the past two decades, an increasing number of cardiac models have been developed at different physical scales such as cells, tissues, and the whole heart. This rapid development is partly due to the improvements in experimental techniques, e.g., whole-cell patch clamp, optical mapping, tomography imaging, and partly to the advanced computing technology. With the proliferation of detailed experimental data, many existing models elaborated various aspects of cardiac functions, e.g., metabolism, electrical activity, mechanical activity, or anatomy.

However, there is only a sparse set of literature relevant to the modeling of spatiotemporal disease processes across the physical scales of increasing complexity, from molecule to cell to tissues to the whole heart. The research communities, including physiology, cardiology, and engineering, have identified the urgent need to model the cardiac system at more than one biological scale. Indeed, multi-scale modeling offers an unprecedented opportunity to link molecular-level information to the whole-heart behaviors. Such a link across physical scales will bring model development one step closer to predictive medicine.

In addition, corroborating data that connect cellular changes in excitability with the whole-heart electrical conduction is difficult without performing computer simulations. Multi-scale cardiac models provide the capability of computer experimentation when physical experimentation is impossible. Computer models can also be used to perform experiments before expensive physical experiments are undertaken. Further, the results from simulation experiments will be able to derive new hypotheses and suggest new designs of physical experiments in our ongoing and future research.

However, computer models of cardiac systems are computationally very expensive. For example, one simulation of the whole heart can take up to 2 weeks in a 3.2 GHz desktop computer. This limits the utility of the models in real-time medical decision making, where the models need to be optimized over a potentially large number of decision variables. Our previous works demonstrated an approach to tackle this computational challenge by approximating the complex cardiac models using statistical metamodels [5]. Although this approach has been widely used in engineering, we have not seen its applications in cardiovascular systems. Because the decision space has physical constraints due to the heart geometry and because the outputs are functional responses, new and/or improved statistical methods need to be developed for this specific application [6]. Furthermore, the models need to be calibrated using real-world data so that we have a greater confidence in using them in practice.

1.4 Summary

This book contributes a rigorous methodological basis about the sensing, modeling, and optimization of cardiac systems. This contribution is significant, because it is conducive to achieving the fundamental understanding of spatiotemporal cardiac disease processes that is so vitally needed to improve preventive healthcare services.

This book will usher in a paradigm shift in healthcare, i.e., from reactive care to preventive and proactive care, from experience-based to evidence-based medical decision making. Further, an early detection of cardiovascular diseases will decrease mortality rates, promote the timely delivery of life-saving interventions, and reduce healthcare cost (e.g., preventive care in lieu of expensive surgical interventions). This book integrates engineering methods and tools (e.g., modeling, simulation, experiments, statistics, and optimization) with biomedical sciences for solving an important health-related problem, which can make a lasting impact in our society. In addition, new simulation models can serve as training tools for students of medicine to improve their understanding of disease progression over time and the correlation with cardiac electrical signals. On the other hand, engineering students will gain a better understanding of healthcare challenge and opportunities when dealing with sensing, modeling, and simulation of cardiac systems.

References

1. C.W. Tsao, A.W. Aday, Z.I. Almarzooq, et al., Heart disease and stroke statistics—2022 update: a report from the American Heart Association. Circulation **145**(8), e153–e639 (2022)
2. S. Nattel, A. Maguy, S. Le Bouter, Y.-H. Yeh, Arrhythmogenic ion-channel remodeling in the heart: heart failure, myocardial infarction, and atrial fibrillation. Physiol. Rev. **87**(2), 425–456 (2007)
3. D. Dubin, *Rapid Interpretation of EKG's: An Interactive Course* (Cover Publishing Company, New York, 2000)
4. H. Yang, Y. Chen, F.M. Leonelli, Whole heart modeling—spatiotemporal dynamics of electrical wave conduction and propagation, in *2016 38th Annual International Conference of the IEEE Engineering in Medicine and Biology Society (EMBC)* (IEEE, New York, 2016), pp. 5575–5578
5. D. Du, H. Yang, A.R. Ednie, E.S. Bennett, Statistical metamodeling and sequential design of computer experiments to model glyco-altered gating of sodium channels in cardiac myocytes. IEEE J. Biomed. Health Inform. **20**(5), 1439–1452 (2015)
6. M. Plumlee, V.R. Joseph, H. Yang, Calibrating functional parameters in the ion channel models of cardiac cells. J. Am. Stat. Assoc. **111**(514), 500–509 (2016)

Chapter 2
Multi-scale Simulation Modeling of Cardiac Systems

Simulation modeling methods inherently provide a fundamental infrastructure for understanding and predicting disease-altered biological processes and treatment outcomes. Undoubtedly, multi-scale integration of simulation models with medical data and in silico experiments is urgently needed and widely recognized by engineers and biomedical scientists. This chapter focuses on the development of multi-scale computer models that can be integrated with physical experiments (e.g., in vivo, in vitro studies) to improve the understanding of disease-altered cardiac electrical dynamics from cells to tissues to the whole heart. This chapter will present computer models of cardiac ion channels, cells, tissues, and the whole heart with realistic geometries for the simulation of the excitation and propagation of cardiac electrical activities.

2.1 Computer Modeling of Ion Channels and Tissues

Computational modeling of the heart provides a fundamental infrastructure for understanding the dynamic cardiac processes under different heart conditions. Different computer models have been developed to investigate the cardiac cell electrophysiology and ion channel dynamics [1–4]. In addition, various simulation models at the tissue level have been further constructed to understand the microstructure and electrophysiology of the cardiac tissue [5–7]. Our prior work developed computer models to show that changes in the glycosylation (i.e., sugar structure) of hERG (the human Ether-à-go-go Related Gene) channels modulate a repolarizing cardiac K+ current, thereby impacting the AP of human ventricular myocytes [3, 8, 9]. Cardiac cell AP is composed of 5 distinct phases (phases 0 to 4; see Fig. 2.1). In phase 4, a cardiac myocyte is in the resting state (−90 mv). When it is stimulated by neighboring cells or a stimulus electric current, the cell quickly depolarizes and opens Na+ channels that bring a rapid influx of Na+ ions (I_{Na}) into

H. Yang, B. Yao, *Sensing, Modeling and Optimization of Cardiac Systems*, SpringerBriefs in Service Science, https://doi.org/10.1007/978-3-031-35952-1_2

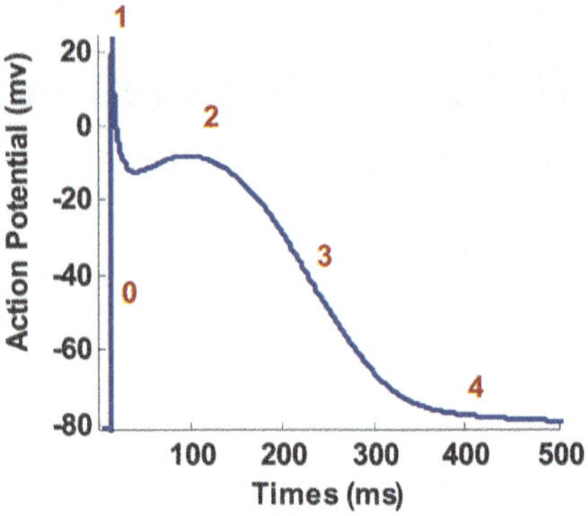

Fig. 2.1 Illustration of 5 phases in the action potential (AP) of cardiac cells

the cell. As shown in Fig. 2.1, the AP rises from the resting potential (−90 mv) to positive voltages (+20 mv) (i.e., phase 0, also called rapid depolarization phase). Then, there is a small downward deflection in phase 1 that is caused by the inactivation of the fast Na+ channels and transient net outward K+ currents (I_{to}). This phase 1 is followed by a plateau AP phase 2 that is sustained mainly by a balance between the outward movement of K+ and the inward movement of Ca++(I_{Ca}). Phase 2 also consists of activities from other electrogenic transport proteins, e.g., the sodium–potassium pump current ($I_{Na,K}$) and the sodium–calcium exchanger current ($I_{Na,Ca}$). Phase 3 is called the rapid repolarization phase, during which the potassium K+ currents (e.g., I_{Kr} currents through hERG channels) get the cell back to the resting potential.

Ion channels are heavily glycosylated in cardiac cells. There are more than 30% of a mature protein's mass comprised of glycan structures [10]. Our previous works showed that changes in glycosylation can influence the gating kinetics of voltage-gated Na+ and K+ channels [11, 12]. Specifically, hERG ion channels are the targets in the pharmaceutical studies for the design of anti-arrhythmic drugs and many LQTS (Long QT Syndrom) medications. We investigated how reduced glycosylation modulates the I_{Kr} currents and hERG ion channels using the equations:

$$
\begin{aligned}
I_{kr} &= G_{kr}\sqrt{\frac{K_O}{5.4}}X_{r1}X_{r2}(V - E_K) \\
X_{r1\infty} &= \frac{1}{1 + \exp -\frac{(V-V_a)}{K_a}} \\
X_{r2\infty} &= \frac{1}{1 + \exp -\frac{(V-V_i)}{K_i}} \\
\frac{dX_{ri}}{dt} &= \frac{X_{ri\infty} - X_{ri}}{\tau_{xri}} \quad i = 1, 2,
\end{aligned}
\tag{2.1}
$$

where G_{Kr} is the cell conductance (nS/pF), E_K is the reversal potential, K_a and K_i are the slope factors, V_i is the voltage of half-inactivation, V_a is the voltage of half-activation, $X_{r1\infty}$ is the steady-state activation, $X_{r2\infty}$ is the steady-state inactivation, τ_{xr1} is the activation time constant, and τ_{xr2} is the inactivation time constant.

The whole-cell patch clamp was used to measure the gating parameters (i.e., half-activation voltage V_a, half-inactivation voltage V_i, and slope factors K_a, K_i) of hERG channels under four conditions of glycosylation (i.e., full glycosylation, reduced sialylation, mannose-rich, and N-Glycanase treated). A total of 44 human ventricular cells were used in the physical experiments. These measured changes of gating parameters were integrated into the kinetic equations of hERG channels, thereby predicting the multi-scale glycosylation effects on cells and tissues. The cardiac cell is analogous to a structured network of electric circuits connected among capacitors, resistors, and batteries. Tusscher–Panfilov model [13, 14] is commonly used to represent the electrophysiological behaviors of ventricular cells from human subjects. The action potential (V) is determined by the following differential equations:

$$
\begin{aligned}
-C_m\frac{dV}{dt} &= I_{to} + I_{Ks} + I_{Kr} + I_{K1} + I_{NaCa} + I_{NaK} + I_{pK} + I_{pCa} + I_{bCa} \\
&\quad + I_{bNa} + I_{Na} + I_{CaL} + I_{stim},
\end{aligned} \tag{2.2}
$$

where t is the time, C_m is the cell capacitance per unit surface area, and I_{stim} is the external stimulus current. The transmembrane currents include transient outward current (I_{to}), slow delayed rectifier K+ current (I_{Ks}), rapid delayed rectifier K+ current (I_{Kr}), inward rectifier current (I_{K1}), Na+/Ca++ exchange current (I_{NaCa}), pump current (I_{NaK}), plateau currents (I_{pK}, I_{pCa}), background currents (I_{bCa}, I_{bNa}), fast Na+ current (I_{Na}), and L-type Ca++ current (I_{CaL}). As shown in Fig. 2.2, our modeling results showed that reduced glycosylation will act to shorten the repolarization period of cardiac Aps, and the rapid delayed rectifier K+ current

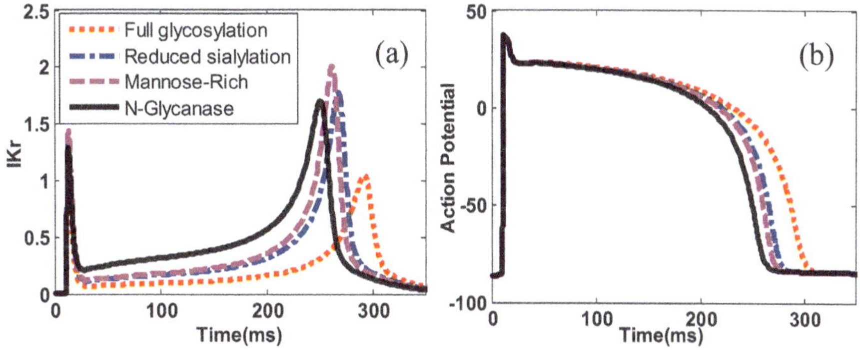

Fig. 2.2 Reduced hERG glycosylation is predicted to alter (**a**) rapid delayed rectifier K+ current (I_{Kr}) and shorten the repolarization of (**b**) action potentials [3, 8]

I_{Kr} is shifted leftward along the time axis with higher current densities (during the action potential).

However, the cardiac cell is not an independent unit but rather is connected to adjacent cells. Each depolarized cell can stimulate neighboring cells and trigger cell-to-cell conductions [8]. In the rapid depolarization phase, the overshoot of Na+ ions causes a depolarizing-to-resting Na+ gradient and drives the flow of Na+ currents through the connexon (i.e., a gap junction channel between two adjacent cells) into the resting cells. The influx of Na+ currents causes neighboring cells to reach the threshold potential and activate the rapid depolarization phase (0) of the AP. As the depolarization and repolarization propagate among cells, electrical waves are generated. A linear strand of cells (i.e., cable or ring) is modeled using the following partial differential equation:

$$\frac{\partial V}{\partial t} = \frac{I_{ion} + I_{stim}}{C_m} + \frac{1}{\rho_x S_x C_m} \frac{\partial^2 V}{\partial x^2}, \tag{2.3}$$

where I_{ion} is the sum of transmembrane ion currents (i.e., as shown in the right hand of Eq. (2.2)), ρ_x is the cellular resistivity, S_x is the surface-to-volume ratio, and C_m is the cell capacitance per unit surface area.

In the computer experiments, we connected ventricular myocytes in a linear or circular fiber, i.e., cable or ring, to predict the glycosylation effects on AP propagation [8]. We conclude from the data in Fig. 2.3 that glycosylation affects the AP propagation in inhomogeneous tissues. As shown in Fig. 2.3a–c, the inhomogeneous cable contains 100 cells, in which half of cells are fully glycosylated and the remaining 50 cells are under one of the three independent conditions of reduced glycosylation (i.e., reduced sialylation (a), mannose-rich (b), or N-Glycanase (c)). The stimulus is initiated in the first cell, and then neighboring cells are consecutively excited until electrical waves are propagated and conducted to the end of the tissue. The abruptions in electrical waves indicate the heterogeneity of APD between fully glycosylated cells and less glycosylated cells. The repolarization heterogeneity potentially increases the risks of cardiac arrythmia. The comparisons are also presented in Fig. 2.3d–f, which represent the differences between three inhomogeneous cables (see Fig. 2.3a–c) and a homogeneous cable, i.e., full glycosylation. The N-Glycanase treated cable is shown to have the most significant abruptions in electrical wave conductions. Similar results were shown in the cell-ring experiments. That is, reduced glycosylation shortens the repolarization duration, and N-Glycanase treated cells return to resting potentials faster than the others.

Furthermore, the electrical conductions on 2D cardiac tissues were modeled with the monodomain reaction–diffusion equations [8]:

$$\frac{\partial V}{\partial t} = \frac{I_{ion} + I_{stim}}{C_m} + \frac{1}{\rho_x S_x C_m} \frac{\partial^2 V}{\partial x^2} + + \frac{1}{\rho_y S_y C_m} \frac{\partial^2 V}{\partial y^2}, \tag{2.4}$$

where I_{ion} is the sum of all transmembrane ionic currents, ρ_x and ρ_y are cellular resistivities, S_x and S_y are surface-to-volume ratios, and C_m is the cell capacitance

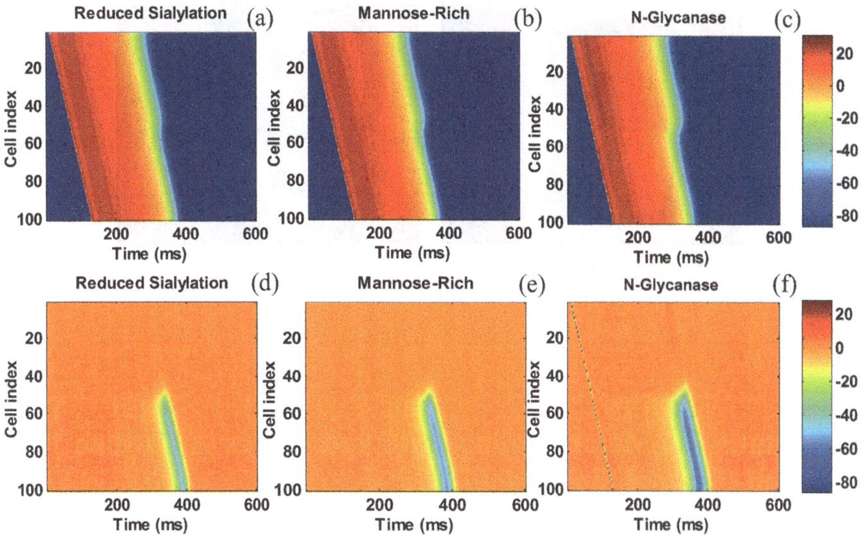

Fig. 2.3 AP propagation along a 1D inhomogeneous cable with 100 cells, in which the first 50 cells are under conditions of full glycosylation and with the second 50 under the following reduced glycosylation conditions: (**a**) reduced sialylation, (**b**) mannose-rich, and (**c**) N-Glycanase. The comparison plots (**d**)–(**f**) represent the differences between (**a**)–(**c**) vs. a cable of full glycosylation [3, 8]

per unit surface area. As shown in Fig. 2.4, reentry waves were simulated in 2D sheets of ventricular tissue (600×600) with the S1-S2 protocol [3, 8]. N-Glycanase treated tissues show a smaller width of waves than the controls, i.e., full glycosylation. This indicates that the vulnerability to arrhythmia is different for the treated tissues, because the width of waves impacts not only electrical conductions in space but also in the time period for rotating waves to bounce back. The theoretical ECG signal was also reconstructed to evaluate whether and how altered glycosylation affects the system-level view of cardiac electrical activity. The ECG derivation assumes an infinite volume conductor and calculates the dipole source density of the membrane potential V in all node points of the medium using the following equation:

$$\mathrm{ECG} = \int_A \frac{D\nabla V \cdot \vec{r}}{r^3} dA, \tag{2.5}$$

where A is the area (for the two-dimensional tissue sheets) or the volume (for the atria, ventricles or the whole heart) over which the ECG is integrated, and r is the vector from the recording electrode to a point in the excitable media. By appropriately choosing the locations, multiple-lead ECG was derived to quantify the variations of cardiac electrical activity under four glycosylation conditions.

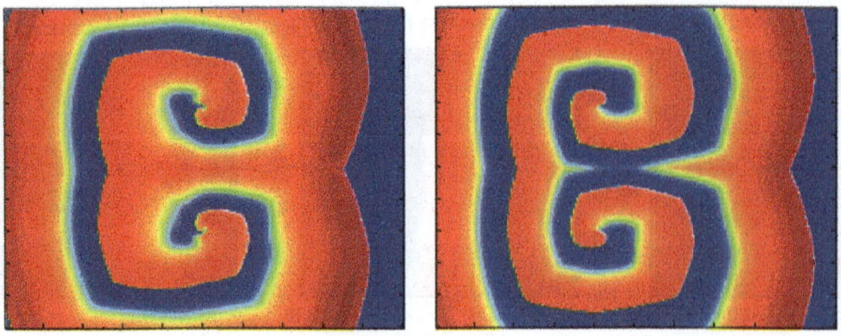

Fig. 2.4 Reduced hERG channel glycosylation is predicted to alter reentry patterns in 2D ventricle simulation [3, 8]

2.2 Statistical Metamodeling and Experiments in Cardiac Ion Channel Simulation

Connecting congenital modifications to high-level system behaviors using multi-scale in silico experiments overcomes many practical limitations associated with using only functional/biochemical studies to be determinative of mechanisms among scales (organizational levels). Computer models can also be used to perform predictive simulations before expensive experiments are undertaken.

Biological or in silico experiments investigate the influence of factors (e.g., pathways, levels of reduced glycosylation) on the response of a biological system. Experimental design focuses on an optimal plan to conduct experiments. In the practice, it is not uncommon that the "one-factor-at-a-time (OFAT)" approach is used to investigate each factor separately, fixing the other factors. Experimenters in medical sciences often have good intuition about the problem when thinking in the OFAT way. Although this is effective to a certain level, this OFAT design requires more experimental runs to achieve precision and also cannot estimate interactions between factors. As such, factorial design or fractional factorial design is increasingly adopted to improve the efficiency and effectiveness of biological experiments. However, a significant gap remains in utilizing in silico experiments to accelerate knowledge discovery in biological experiments.

In silico models of cardiac systems involve a large set of nonlinear differential equations. Nonlinearity poses significant challenges for the design of in silico experiments. It is also impractical to use a large number of experimental runs because solving nonlinear differential equations is time-consuming. There is an urgent need to optimize experimental effectiveness and minimize the number of runs (i.e., reduce the computational cost and time) for in silico experiments. However, very little has been done to develop optimal design of in silico experiments and then exploit the acquired knowledge to identify the effects of regulated glycosylation on specific electromechanical proteins in biological experiments.

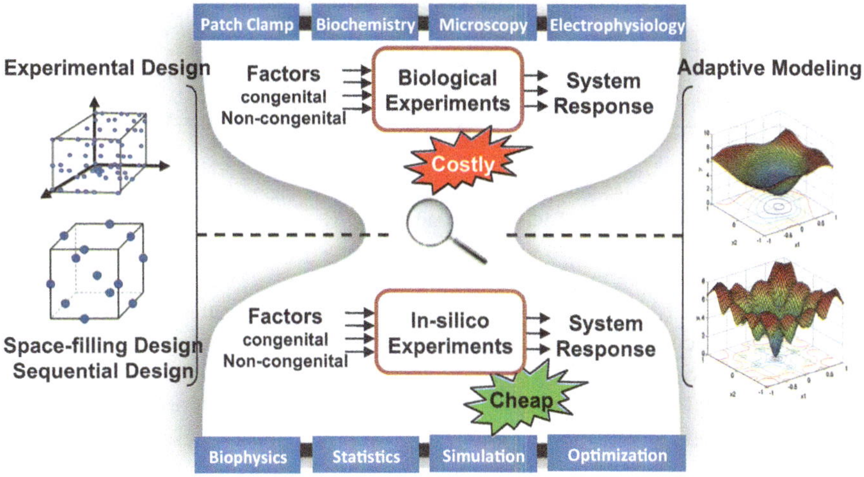

Fig. 2.5 Integration of in silico and biological experiments for knowledge discovery

This section presents statistical design of in silico experiments on simulation models for testing the hypothesis before biological experiments are undertaken. Such in silico results can be used to suggest new biological experiments to question whether and how specific pathways/proteins are altered by regulated glycosylation. As shown in Fig. 2.5, there is a significant gap between biological experiments and in silico ones. There are many practical and ethical limitations to perform biological experiments in multiple organization levels. Also, it is very expensive and difficult to comprehensively investigate factor–response relationships across biological scales. On the other end, many in silico models are developed to only corroborate data obtained from biological experiments. Very little has been done to design and analyze in silico experiments and then integrate the results with biological experiments to accelerate knowledge discovery.

Cardiac models involve a large set of nonlinear equations with a high-dimensional space of factors of interest. The first step is to screen these factors so as to identify important factors sensitive to the response function. For example, if there are a set of 25 factors of interest, a full factorial design would require $2^{25} = 33{,}554{,}432$ runs, but a fractional factorial design needs only 1024 runs (2_V^{25-15}). Based on the effect hierarchy principle, the resolution V ensures that main pathway effects are strongly clear and two-pathway interactions are clear. As such, the design and analysis of in silico experiments are indispensable for factor screening. This will greatly reduce the dimensionality of factor space, thereby improving the effectiveness and efficiency of adaptive modeling.

Furthermore, our prior work developed a new statistical metamodeling approach for the sequential design of in silico experiments and the calibration of models of Na+ channels in cardiac cells. This approach is implemented and evaluated with real-world in vitro experimental data of glyco-altered gating kinetics in the

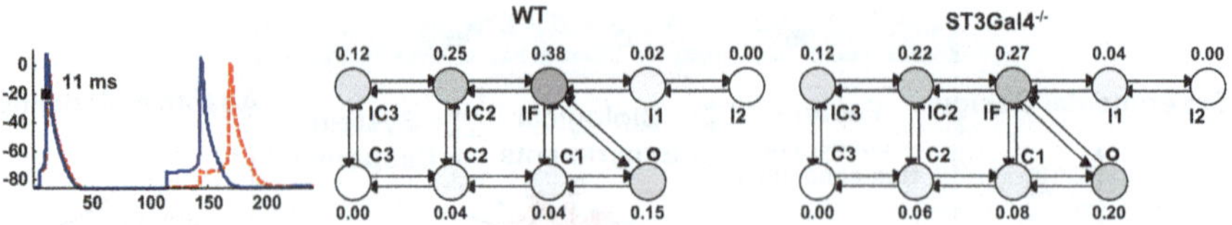

Fig. 2.6 Markov state transitions and distributions of sodium channels during the AP upstroke under both wild-type (WT) and ST3Gal4$^{-/-}$ conditions [15]

ST3Gal4$^{-/-}$ heart [15]. In the literature of biological experiments, this is a new methodological contribution for calibrating large-scale computer models. Statistical design of experiments is an open-box methodology for calibrating multi-scale cardiac models. In contrast, traditional methods (e.g., genetic algorithms, evolution strategies, particle swarm optimization) are black-box approaches, which can provide perfect fits for some time but reveal little information on the underlying physics. These traditional methods work well when facing a low-dimensional input vector but encounter significant challenges when dealing with high-dimensional functional responses (e.g., AP curves). Design of in silico experiments is conducive to providing open-box knowledge about electrophysiological physics, while improving the computational efficiency by the reduction of the number of experimental runs (costs).

Markov modeling of ion channels under conditions of reduced glycosylation poses significant challenges because of nonlinear, non-convex, and ill-posed characteristics. We developed new algorithms to calibrate the Markov model and estimate the optimal model parameters to simulate the gating kinetics of ion channels in altered glycosylation and in control. Figure 2.6 shows the variations of mechanistic details in the state transitions and distributions in the Markov model of Na+ ion Channels for both wild-type (WT) and ST3Gal4$^{-/-}$ myocytes. Note that in vitro experiments can measure ionic currents and AP but preclude us from determining directly the change in transitions among molecular states. For instance, the Markov model of Na+ channels includes 1 open state (O), 1 fast inactivation (IF), 2 intermediate inactivated states (I1, I2), 3 close states (C1, C2, C3), and 2 closed-inactivation states (IC2, IC3). In silico modeling provides greater flexibility than in vitro experiments to examine the molecular states during the time course of AP. For example, Fig. 2.6 shows ST3Gal4$^{-/-}$ channels open slightly faster and wider than WT (i.e., 20% vs. 15%) during the AP upstroke. This also leads to a shorter time-to-peak value for ST3Gal4$^{-/-}$ channels.

Simulation modeling at the cellular and tissue level plays an important role in understanding the microstructure of cardiac processes, but they are limited in the ability to quantitatively elucidate the whole-heart electrical activity and simulate disease-altered ECG signals. As such, they are limited in the ability to serve as a decision-planning tool in clinical practice. Mathematical models based on differential equations have also been developed to generate body-surface full-cycle ECG signals. For example, Quiroz-Juarez et al. [16, 17] proposed an extended heterogeneous oscillator model to characterize the cardiac conduction system and further generated 12-lead ECG waveforms by numerically solving the oscillator model. However, such oscillator models generate ECG signals without modeling the whole-heart electrodynamics, which are not capable of quantitatively investigating the heart dynamic function and disease-altered tissue heterogeneity. In the next section, we will introduce our work on whole-heart modeling to simulate the spatiotemporal dynamics under different cardiac conditions, which can be further leveraged to design decision-planning tools for precision diagnosis and treatment of heart disease in clinical applications.

2.3 Whole-Heart Computer Simulation

Heart rhythm is generated by the excitation and propagation of electrical signals across different heart chambers. Specifically, as shown in Fig. 2.7, a cardiac cycle starts with the activation of the sinoatrial node (SAN) and spreads out through the right atrium (RA) and left atrium (LA). The electrical wave will then arrive at the atrioventricular node (AVN) and travel through the bundle of His (His) toward Purkinje (PKJ) fibers after both atria completely depolarize, leading to electrical depolarization and repolarization of the right ventricle (RV) and left ventricle (LV) to complete the cycle [18]. Cardiac electrodynamics travels from the heart to the body surface, and the resulted body-surface electrical signals are characterized by the ECG waveform, which generally consists of a P wave, QRS complex, and a T wave. In order to fully model the whole-heart electrical activity and generate the full-cycle ECG signals, the electrical conduction path and electrophysiological heterogeneity in different heart regions need to be incorporated into the computer simulation.

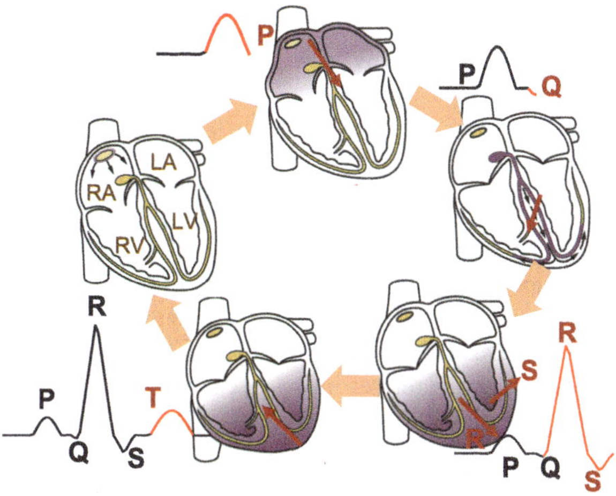

Fig. 2.7 Cardiac depolarization–repolarization cycle and the corresponding segment in the ECG signal (adapted from [18])

Recently, different whole-heart models have been developed to study disease-altered cardiac electrodynamics. For example, Balakrishnan et al. [19] developed a 2D whole-heart model to investigate cardiac arrhythmia. They constructed a 2D geometry by including key components of the electrical conduction system and simulated the cardiac electrodynamics by numerically solving the FitzHugh–Nagumo model [20] across the 2D geometry. Solvilj et al. [21] integrated the cardiac conduction system into a simplified 3D ellipsoid heart model to simulate the whole-heart electrical activity and generate full ECG cycles. Schenone et al. [22] simulated cardiac cycles on a realistic heart geometry under healthy and bundle-block cardiac conditions. They modeled the cardiac electrodynamics in atria and ventricles separately using two different ionic models and added a delay manually between the atria and ventricular excitation, which might not be effective to simulate the interaction between the electrodynamics in atria and ventricles and cannot be used to simulate the electrical activities of the heart under atrial fibrillations. Most existing whole-heart simulation models did not incorporate the conduction path of the human heart into cardiac simulation or failed to consider the refractory heterogeneous during the ventricular repolarization in an anatomically detailed heart geometry. In this case study, we develop a whole-heart model to simulate the spatiotemporal cardiac electrodynamics on an anatomically detailed geometry and further construct full ECG cycles on the body surface.

Aliev–Panfilov Model

In the literature, the two-variable Aliev–Panfilov (A-P) model [23–26] has been widely utilized to describe the spatiotemporal propagation of electrodynamics in the heart:

$$\frac{\partial u}{\partial t} = \nabla \cdot (D\nabla u) + f(u, v), \quad \frac{\partial v}{\partial t} = g(u, v), \tag{2.6}$$

where u denotes the normalized transmembrane potential, v models the recovery behavior of cardiac cells, and D is the conductivity that determines the cardiac conduction velocity. Moreover, the reaction part, i.e., $f(u, v)$ and $g(u, v)$, of the model is represented by

$$f(u, v) = ku(u - a)(1 - u) - uv \tag{2.7}$$

$$g(u, v) = \xi(u, v)(-v - ku(u - a - 1)), \tag{2.8}$$

where parameters k, a, and ξ characterize the cardiac tissue property and define the shape of the action potential (i.e., time course of the transmembrane potential u). Specifically, parameters k and a control the repolarization and excitability of cardiac cells, respectively; $\xi(u, v) = \xi_0 + \mu_1 v/(u + \mu_2)$ defines the coupling strength between u and v; refractoriness ξ_0 characterizes the action potential duration (APD), and bigger ξ_0 will result in shorter APD.

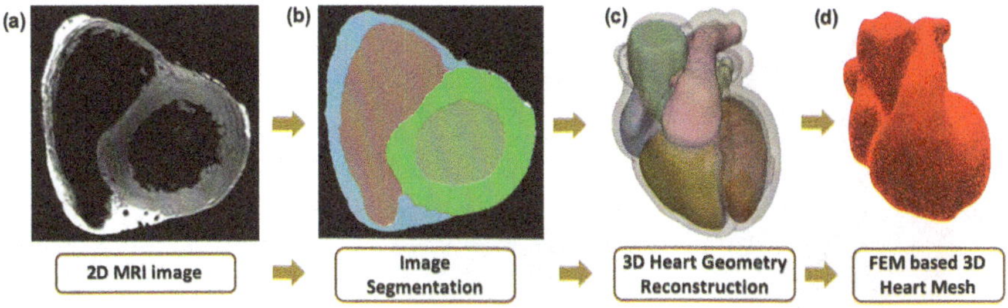

Fig. 2.8 The procedure to reconstruct the 3D heart geometry. (**a**) 2D MRI image; (**b**) Image segmentation; (**c**) Reconstructed 3D geometry of the heart; (**d**) Discretized heart mesh from finite element method (FEM)

Heart Geometry Reconstruction

Accurate and robust reconstruction of the 3D heart geometry is crucial to simulation modeling of whole-heart electrophysiology. The rapid advancements in image sensing such as magnetic resonance imaging (MRI) and computed tomography (CT) have enabled the noninvasive visualization of internal anatomical cardiac structures and functionality, greatly facilitating the reconstruction of heart geometry to support treatment design and surgical planning for heart disease. We reconstruct the anatomically realistic heart geometry based on 2D medical images provided by the Visible Human Project from the National Library of Medicine [27]. The reconstruction procedure is conducted using 3D slicer [28] and Meshmixer,[1] which are widely used software for 3D geometry reconstruction and mesh optimization. As shown in Fig. 2.8, the 2D MRI images (illustrated by Fig. 2.8a) are first segmented and labeled to represent different heart regions (see Fig. 2.8b). Second, the segmented images are stacked together and interpolated to reconstruct the 3D geometry (Fig. 2.8c). Finally, the 3D geometry is discretized into heart mesh from finite element analysis as shown in Fig. 2.8d, and the discretized mesh is further optimized by the Meshmixer software for reliable numerical simulation of the cardiac electrodynamics.

In this case study, we focus on the electrical activities on the four heart chambers, i.e., RA, LA, RV, and LV, as shown in Fig. 2.9a. The atria geometry is modeled as a discretized triangular surface with 3801 nodes and 7525 triangle elements. In order to account for the electrophysiological heterogeneity among the epicardium, mid-myocardium, and endocardium of the ventricles, the ventricle geometry is discretized into a tetrahedron mesh with 13,590 nodes and 54,165 mesh elements. Furthermore, to fully incorporate the cardiac electrical conduction path, we have modeled the AVN, His, BNL, and PKJ as shown in Fig. 2.9b, in which different electrophysiological parameters will be specified to fully model the conduction pattern of cardiac electrodynamics and replicate the cardiac cycle. Table 2.1 shows

[1] Autodesk, Inc. Meshmixer, 2016. http://www.meshmixer.com/.

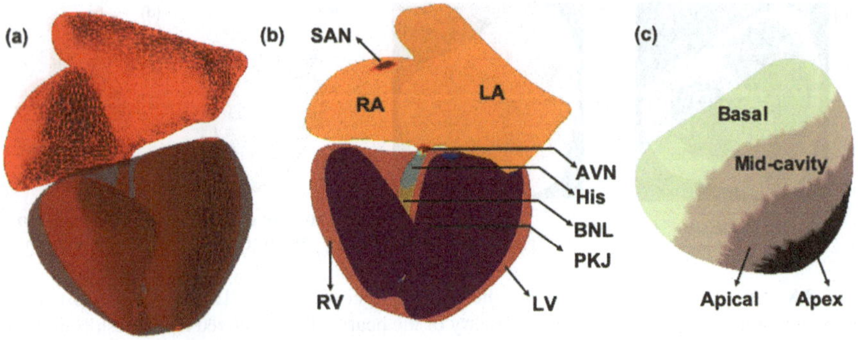

Fig. 2.9 (**a**) Discretized geometry of the four chambers with 21,008 mesh elements; (**b**) The electrical conduction system of the heart; (**c**) Ventricular segmentation according to the activation time

Table 2.1 Parameter setting for whole-heart simulation

	$D(mS \cdot m^{-1})$	k	a	ξ_0	μ_1	μ_2
RA & LA	8	8	0.12	0.002	0.2	0.3
AVN	0.5	8	0.15	0.002	0.2	0.3
His	10	8	0.15	0.002	0.2	0.3
BNL	15	8	0.15	0.002	0.2	0.3
PKJ	35	8	0.10	0.002	0.2	0.3
Ven_Apex (Epi)	8	8	0.10	0.004	0.2	0.3
Ven_Apex (Mid-myo)	8	8	0.10	0.002	0.2	0.3
Ven_Apex (Endo)	8	8	0.10	0.001	0.2	0.3
Ven_Apical (Epi)	8	8	0.10	0.008	0.2	0.3
Ven_Apical (Mid-myo)	8	8	0.10	0.004	0.2	0.3
Ven_Apical (Endo)	8	8	0.10	0.002	0.2	0.3
Ven_Mid-cavity (Epi)	8	8	0.10	0.012	0.2	0.3
Ven_Mid-cavity (Mid-myo)	8	8	0.10	0.006	0.2	0.3
Ven_Mid-cavity (Endo)	8	8	0.10	0.003	0.2	0.3
Ven_Basal (Epi)	8	8	0.10	0.016	0.2	0.3
Ven_Basal (Mid-myo)	8	8	0.10	0.008	0.2	0.3
Ven_Basal (Endo)	8	8	0.10	0.004	0.2	0.3

the parameter settings for the A-P model in different heart regions. Homogeneous parameters are assigned to the atria, while the ventricles are divided into subsegments with heterogeneous parameter settings for the refractoriness (i.e., ξ_0). This is to generate the correct T wave, which will be detailed in the next subsection. The conductivity setting (i.e., D) is taken from the existing literature [21, 29]. Note that a small conductivity (i.e., $0.5\,\mathrm{mS} \cdot \mathrm{m}^{-1}$) is assigned to AVN to guarantee that both atria are fully depolarized before the ventricles start to depolarize. The values for parameters k, a, ξ_0, μ_1, and μ_2 in the A-P model (i.e., Eq. (1)–(3)) are selected to regulate the shape of action potentials in different heart regions. Note that here we set the parameter values according to the general assumptions for the electrophysiological property of the human heart, while the true parameter setting for each patient needs to be calibrated based on the personalized medical data [25, 26, 30].

Modeling the Heterogeneity in Ventricular Repolarization

Existing in vivo experiments [31, 32] have demonstrated that the electrical activation in the ventricles first occurs from endocardial apex to epicardial apex, and then from apex to base, while the repolarization occurs from the base to apex and epicardium to endocardium. This is the reason that the T wave (due to ventricular repolarization) has the same polarity with the QRS interval (due to ventricular depolarization). As such, both the transmural and apex-to-base heterogeneities need to be accounted for in order to model the ventricular depolarization and repolarization and further generate realistic full-cycle ECG signals.

In order to capture the apex-to-base heterogeneity, we divide the geometry of each ventricle into 4 segments: apex, apical region, mid-cavity, and basal region according to the geodesic distance between each region to the endocardial apex, as shown in Fig. 2.9c. The geodesic distance between two nodes on a complex 3D surface is defined as the length of the shortest path on the surface connecting the two nodes. A n_k-nearest neighbor graph is first constructed by connecting each location $\boldsymbol{s}_i$ to its n_k-nearest neighbors $\boldsymbol{s}_{j,j\leftrightarrow i}$ (where $j \leftrightarrow i$ means node j is one of the n_k-nearest neighbors for node i or i and j are connected) on the geometry. Second, the distance between each pair of nodes on the graph is initialized as: $d_G(i, j) = \|\boldsymbol{s}_i - \boldsymbol{s}_j\|$ (Note $\|\cdot\|$ denotes the Euclidean norm) if i and j are connected; $d_G(i, j) = \infty$, otherwise. Third, the Dijkstra's algorithm [33] is implemented to calculate the shortest path (the length of which is the geodesic distance) between each node and the endocardial apex, which is further implemented to segment the ventricular geometry into the apex, apical region, mid-cavity, and basal region.

Furthermore, we assign the refractoriness (i.e., ξ_0) of apex, apical region, mid-cavity, and basal region with increasing values (i.e., assigning biggest ξ_0 to the basal region and smallest ξ_0 to the apex). Note that the region with bigger ξ_0 will generate the action potential with shorter APD and will repolarize earlier compared with the smaller ξ_0 area [34]. As such, the apex will be the first to depolarize and the last to repolarize. Additionally, we assign different values of ξ_0 to epicardial, mid-myocardial, and endocardial cells to account for the transmural heterogeneity with a bigger ξ_0 to epicardium and a smaller ξ_0 to endocardium. The detailed parameter setting for different regions in the ventricular geometry is provided in Table 2.1. Figure 2.10 shows the simulated action potentials in different ventricular regions, demonstrating the apex-to-base and transmural heterogeneities.

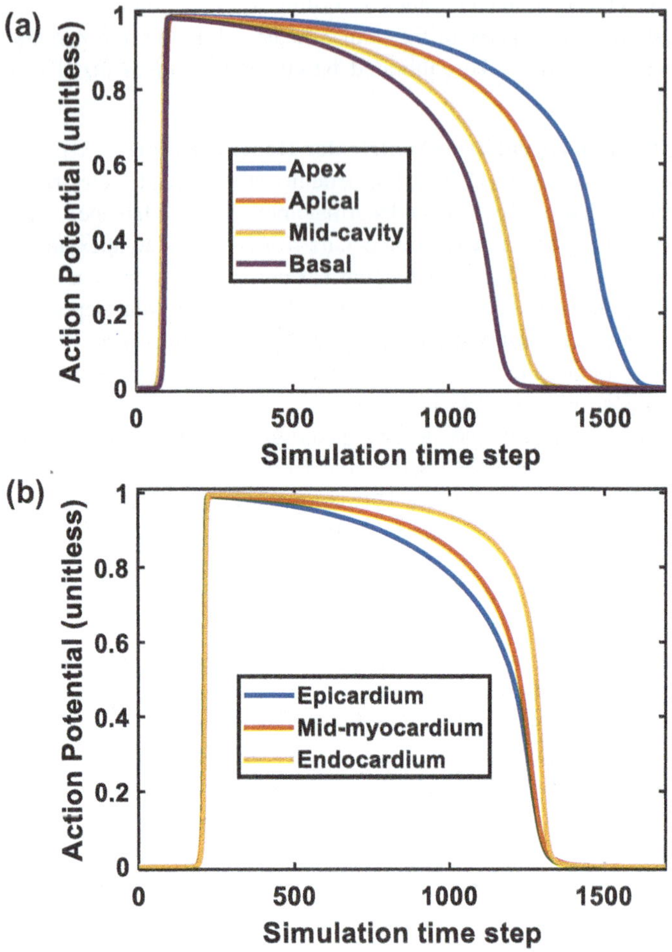

Fig. 2.10 (**a**) Apex-to-base (**b**) Transmural heterogeneity in the shape of action potential

Numerical Simulation of the Whole-Heart Electrodynamics with Finite Element Method

We propose to numerically solve the A-P reaction–diffusion model (i.e., Eq. (1)–(3)) and simulate the cardiac electrodynamics on the 3D heart geometry using finite element method (FEM) with operator splitting [35] including three steps:

Step (1): Integrate the nonlinear reaction part for one half time step (i.e., from t to $t + \delta t/2$) given the initial condition at time t:

$$\frac{\partial u}{\partial t} \approx \frac{u_{t+\delta t/2} - u_t}{\frac{1}{2}\delta t} = f(u_t, v_t) \tag{2.9}$$

$$\frac{\partial v}{\partial t} \approx \frac{v_{t+\delta t/2} - v_t}{\frac{1}{2}\delta t} = g(u_t, v_t). \tag{2.10}$$

Step (2): Consider the results from Step (1) as the initial condition (i.e., $u_{*,t} = u_{t+\delta t/2}$ and $v_{*,t} = v_{t+\delta t/2}$), and integrate the linear diffusion part for a full time step (i.e., from t to $t + \delta t$) as

$$\frac{\partial u}{\partial t} \approx \frac{u_{*,t+\delta t} - u_{*,t}}{\delta t} = \nabla \cdot (D\nabla u_{*,t+\delta t}). \tag{2.11}$$

Solving Eq. (2.11) involves the spatial integration over the complex heart geometry, which is analytically intractable. Here, we utilize the FEM approach to numerically solve the complex integration [36, 37]. Specifically, assume the 3D heart geometry is discretized into a mesh with N nodes and N_T elements, the electric potential at an arbitrary location s is approximated as

$$u(s) = \sum_{i=1}^{N} u_i \phi_i(s), \tag{2.12}$$

where u_i denotes the electric potential at node i and ϕ_i is the linear basis function associated with node i. Multiplying Eq. (2.11) with a linear test function ϕ_j, and integrating it over the heart domain Ω_H, the following result is obtained from the divergence theorem and boundary condition of cardiac electrophysiology [36, 37]:

$$\begin{aligned}
\int_{\Omega_H} \frac{u_{*,t+\delta t}(s) - u_{*,t}(s)}{\delta t} \phi_j dV &= \int_{\Omega_H} \nabla \cdot (D\nabla u_{*,t+\delta t})\phi_j dV \\
&= \int_{\Omega_H} \nabla \cdot (\phi_j D\nabla u_{*,t+\delta t}) dV - \int_{\Omega_H} \nabla\phi_j \cdot (D\nabla u_{*,t+\delta t}) dV \\
&= \int_{S_H} (\phi_j D\nabla u_{*,t+\delta t}) \cdot d\mathbf{S} - \int_{\Omega_H} \nabla\phi_j \cdot (D\nabla u_{*,t+\delta t}) dV \\
&= -\int_{\Omega_H} \nabla\phi_j \cdot (D\nabla u_{*,t+\delta t}(s)) dV.
\end{aligned} \tag{2.13}$$

Plugging Eq. (2.12) into Eq. (2.13) and rearranging it, we have

$$\begin{aligned}
&\sum_{i=1}^{N} u_{(i)*,t+\delta t} \left(\int_{\Omega_H} \phi_i \phi_j dV + \delta t \int_{\Omega_H} (\nabla\phi_i)^T D \nabla\phi_j dV \right) \\
&= \sum_{i=1}^{N} u_{(i)*,t} \left(\int_{\Omega_H} \phi_i \phi_j dV \right),
\end{aligned} \tag{2.14}$$

which can be further written in the matrix form as

$$(\mathbf{M} + \delta t \cdot \mathbf{K})\boldsymbol{u}_{*,t+\delta t} = \mathbf{M}\boldsymbol{u}_{*,t}, \tag{2.15}$$

where $\mathbf{M}$ and $\mathbf{K}$ are the mass matrix and stiffness matrix, respectively, whose element is defined as $\mathbf{M}_{ij} = \int_{\Omega_H} \phi_i \phi_j dV$ and $\mathbf{K}_{ij} = \int_{\Omega_H} (\nabla\phi_i)^T \cdot D\nabla\phi_j dX$. The linear system in Eq. (2.15) can be solved by LU decomposition [38] to increase the numerical stability and computational efficiency.

Step (3): Consider the results given by Step (2) as the initial condition (i.e., $u_{**,t+\delta t/2} = u_{*,t+\delta t}$ and $v_{**,t+\delta t/2} = v_{*,t+\delta t}$), integrate the nonlinear reaction part for another half time step from $t + \delta t/2$ to t, and solve for $u_{t+\delta t}$ and $v_{t+\delta t}$:

$$\frac{u_{t+\delta t} - u_{**,t+\delta t/2}}{\frac{1}{2}\delta t} = f(u_{**,t+\delta t/2}, v_{**,t+\delta t/2}) \tag{2.16}$$

$$\frac{v_{t+\delta t} - v_{**,t+\delta t/2}}{\frac{1}{2}\delta t} = g(u_{**,t+\delta t/2}, v_{**,t+\delta t/2}). \tag{2.17}$$

The propagation of the cardiac electrodynamics will then be obtained by iteratively solving the FEM-based operating splitting procedure given by Steps (1)–(3). The resulted cardiac electrodynamics will then be integrated with the forward ECG modeling to derive the body-surface ECG cycles.

Body-Surface ECG Derivation from Cardiac Simulation

This section describes the derivation of body-surface ECG signals from the simulated cardiac electrodynamics. According to Ohm's law, the electric current density in the heart is computed from the cardiac electric potentials as

$$\boldsymbol{j}_m = -\sigma_H \nabla V_m, \tag{2.18}$$

where σ_H denotes the conductivity of the cardiac tissue, which is proportional to parameter D, and V_m denotes the transmembrane electric potential that is derived from the normalized action potential u, i.e., $V_m \simeq 100u - 80$.

The torso is approximated as a homogeneous volume conductor with the conductivity of σ_B (~0.2 S/m [21]). Then, according to the volume conductor theory, the electric potential at location $\boldsymbol{s}_B$ on the body surface is approximated:

$$\Phi(\boldsymbol{s}_B) = \frac{1}{4\pi\sigma_B} \int_{\Omega_H} \boldsymbol{j}_m \cdot \nabla \left(\frac{1}{\|\boldsymbol{s}_B - \boldsymbol{s}_H\|} \right) d\boldsymbol{s}_H, \tag{2.19}$$

where $\boldsymbol{s}_H$ denotes the coordinates on the heart geometry. Note that the dipole moment of an infinitesimal volume $d\boldsymbol{s}_H$ is defined as $\boldsymbol{p}(\boldsymbol{s}_H) = \boldsymbol{j}_m d\boldsymbol{s}_H = -\sigma_H \nabla V_m(\boldsymbol{s}_H) d\boldsymbol{s}_H$. Approximately, the dipole moment in the volume Ω_k of the cardiac mesh element k is formulated as $\boldsymbol{p}_k = -\sigma_H \Omega_k \nabla V_m(\boldsymbol{s}_k)$. Hence, the integration in Eq. (2.19) can be approximated by a series of summation over the mesh elements as

$$\Phi(s_B) = \frac{1}{4\pi\sigma_B}\sum_k \frac{\sigma_H \Omega_k \nabla V_m(s_k)\cdot(s_k - s_B)}{\|s_k - s_B\|^3}. \tag{2.20}$$

The calculation of gradient on a mesh element, i.e., $\nabla V_m(s_k)$, can be obtained from the Green–Gauss Theorem. Given the body-surface potential mapping in Eq. (2.20), the 12-lead ECG signals can be further derived to quantitatively study the morphology variations under different healthy and diseased heart conditions.

Simulation Results in Healthy Heart

Figure 2.11 shows the excitation and propagation of cardiac electrodynamics in one heart cycle simulated by the proposed model. At the beginning of each cycle, an instantaneous electrical impulse is applied to stimulate the SAN pacemaker

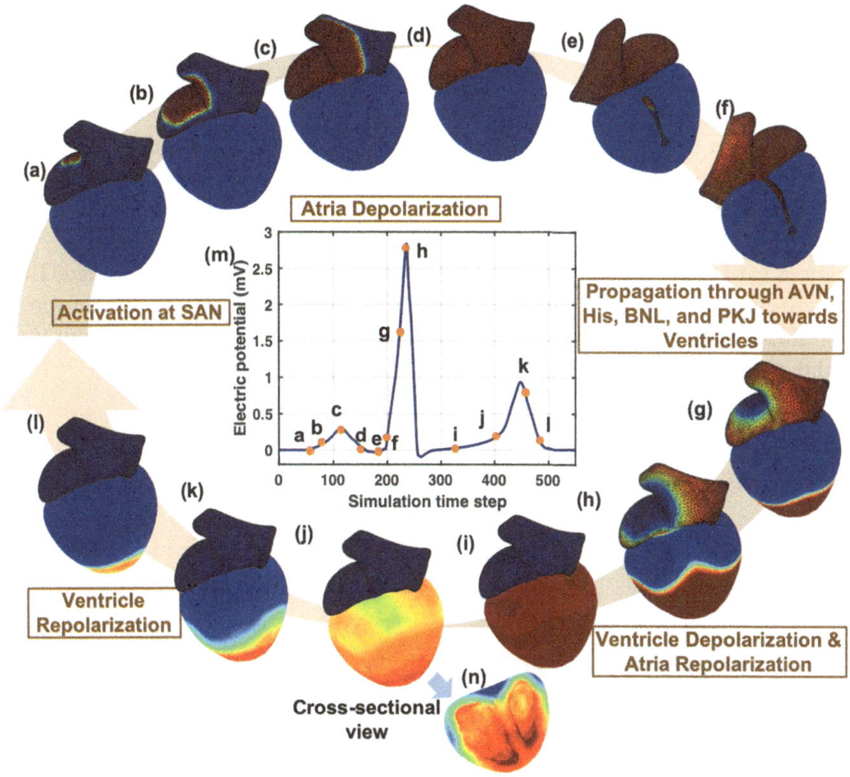

Fig. 2.11 (**a**)–(**l**) Simulated excitation and propagation of electrical waves in one heart cycle; (**m**) the simulated ECG cycle. Note that the labels on the ECG signal denote the time points of activation sequences shown in (**a**)–(**l**). (**n**) The cross-sectional view of electrical potential distribution in the ventricles in (**j**)

and initiate the cardiac electrodynamics (Fig. 2.11a). The electrical impulse then propagates through the atria, leading to atria depolarization as shown in Fig. 2.11b–d. The wavefront is then delayed when reaching the AVN until the entire atria have been activated and depolarized (Fig. 2.11e), which is achieved by assigning a small conductivity (i.e., $D = 0.5\,\text{mS} \cdot \text{m}^{-1}$ as shown in Table 2.1) to AVN. Subsequently, the bundle of His is activated, and the electrical waves further spread through the BNL and PKJ toward the two ventricles (Fig. 2.11f), resulting in ventricular depolarization. Note that atria repolarization also happens during the ventricular depolarization as shown in Fig. 2.11g–i, and thus there is no distinctly visible wave representing atrial repolarization in the ECG cycle [39]. Finally, after the entire ventricles get activated and depolarized, ventricular repolarization starts to occur. The repolarization propagates from the base region, across mid-cavity and apical region, to the apex, and from epicardium, across mid-myocardium, to endocardium as shown in Fig. 2.11i–l and n. The heart cycle completes after the entire ventricles are repolarized. Figure 2.11m illustrates the corresponding ECG signal for one heart cycle with the labels denoting the time points of activation sequences indicated by Fig. 2.11a–l. As illustrated by Fig. 2.11a–m, we have successfully incorporated the electrical conduction system and correctly replicate the full ECG cycle.

We further derive the body-surface ECG signals from the simulated cardiac electrodynamics. As shown in Fig. 2.12a, we model the propagation of electrodynamics from the heart to the body surface through the forward ECG method. In the present investigation, the body surface is triangulated with 352 discretized nodes and 677 mesh elements. Figure 2.12b shows the calculated body-surface potential mapping (BSPM) including 352 ECG channels. Each channel consists of a full ECG cycle composed of the P wave, QRS complex, and T wave. Additionally, the 12-lead ECG is derived from BSPM as shown in Fig. 2.12c, whose lead placement is given by Fig. 2.12a. Specifically, we derive the three Einthoven leads (i.e., V_I, V_{II}, and V_{III}) and the three Goldberger-augmented leads (i.e., aV_R, aV_L, and aV_F) as

$$
\begin{aligned}
V_I &= V_L - V_R, \quad & aV_R &= \frac{2V_R - V_L - V_F}{2} \\
V_{II} &= V_F - V_R, \quad & aV_L &= \frac{2V_L - V_R - V_F}{2} \\
V_{III} &= V_F - V_L, \quad & aV_R &= \frac{2V_F - V_L - V_R}{2}.
\end{aligned}
$$

The six precordial leads, i.e., V_1, V_2, V_3, V_4, V_5, and V_6, are directly obtained from the corresponding BSPM channels. According to Fig. 2.12c, the morphology of simulated 12-ECGs agrees well with the known electrophysiological behavior of the human heart [18].

Whole-Heart Simulation with Myocardial Infarction

Myocardial infarction (MI) occurs due to the blockage of coronary arteries leading to a significant decrease in the blood flow or oxygen supply to the heart and further damaging the heart muscle [40]. The electric conduction in the infarcted area is significantly slower compared with the healthy cardiac muscle [41]. Thus, we set the

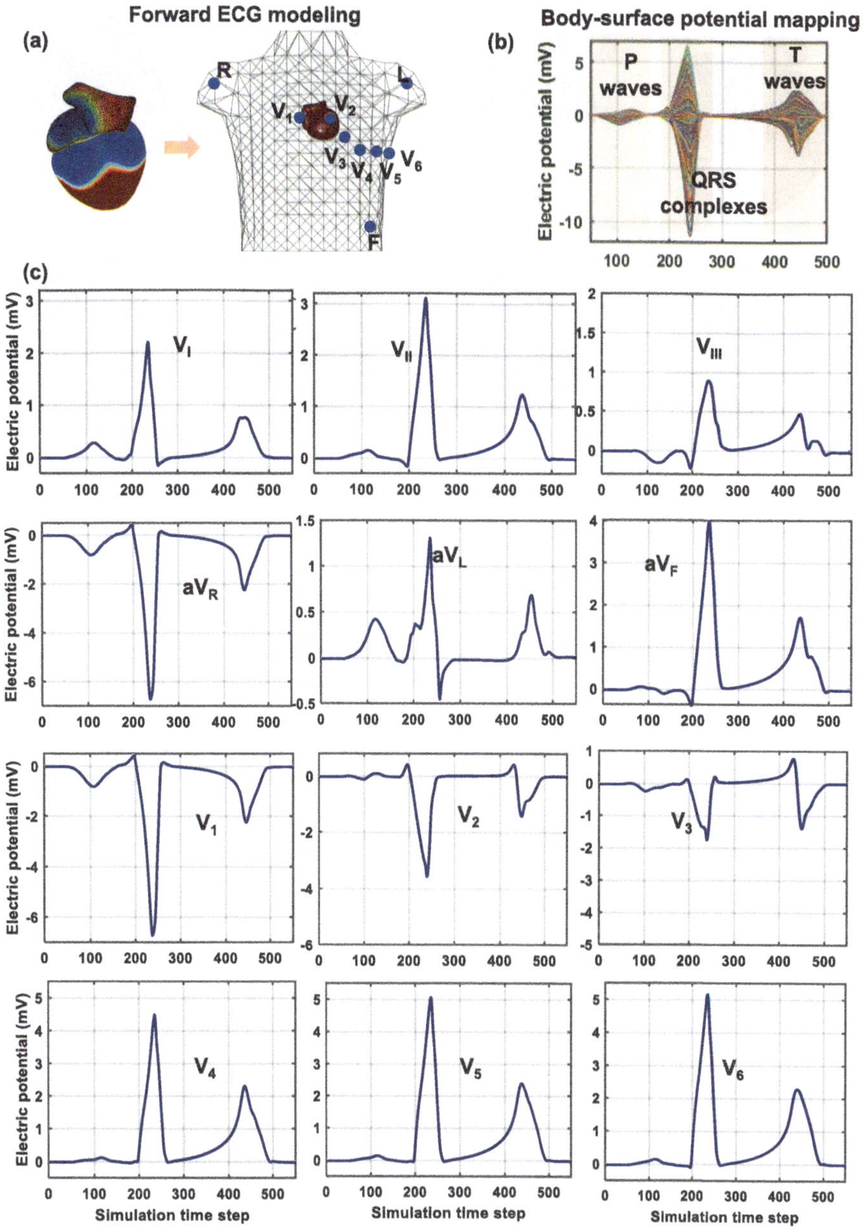

Fig. 2.12 (**a**) The placement of ECG leads; (**b**) Simulated body-surface signals; (**c**) 12-lead ECG signals

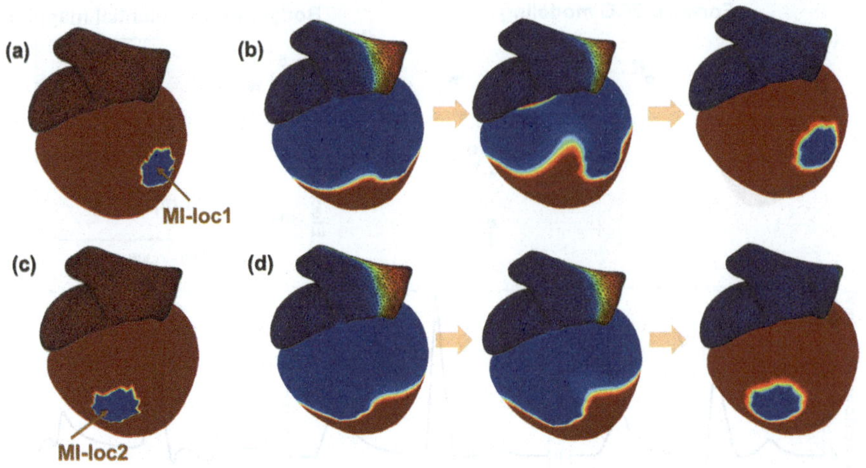

Fig. 2.13 Myocardial infarction occurs at (**a**) location 1 (MI-loc1) and (**c**) location 2 (MI-loc2); Propagation of cardiac electrodynamics with (**b**) MI-loc1 and (**d**) MI-loc2

conductivity in MI regions as zero to simulate the inexcitable MI tissue. Figure 2.13 illustrates the simulated whole-heart electrical activities with MI lesion occurring at different locations, i.e., MI-loc1 and MI-loc2. Note that the lesion region stays inactivated throughout the entire cardiac cycle under both MI conditions.

Figure 2.14 shows the simulated Lead-V_I and Lead-V_2 ECG signals under healthy, MI-loc1, and MI-loc2 cardiac conditions. Note that the P wave signals are identical under the three different cardiac conditions for both ECG leads. This is due to the fact that both MI conditions in the present experiments occur in the ventricles, which will not impact the atria depolarization (i.e., the P wave) for deterministic simulation modeling. Contrarily, the morphology of the QRS complex and T wave is altered due to the impact of MI on both ventricular depolarization and repolarization. Specifically, the Lead-V_I ECG cycle reveals ST-segment depression for both MI-loc1 and MI-loc2, respectively, compared with the healthy ECG cycle as shown in Fig. 2.14a. Similarly, both MI-loc1 and MI-loc2 manifest themselves in the ST-segment depression in the Lead-V_2 ECG as shown in Fig. 2.14b. Note that abnormal ST-segment is a major indicator of MI in clinical practice [42], which is consistent with our simulation results. The proposed cardiac model has successfully simulated MI-altered electrodynamics and has significant potential to quantitatively characterize the location and extent of MI in the heart, which can assist cardiologists in optimal MI diagnosis and effective surgical planning.

Whole-Heart Simulation with Atrial Fibrillation

Atrial fibrillation (AF) is generally initiated by a rapidly firing trigger that often arises in the pulmonary veins [43]. The resulted rapid and uncoordinated activation of the electrical waveforms will lead to abnormal and irregular heart rhythms. In the present investigation, we simulate the AF trigger in the LA (see Fig. 2.15a), which

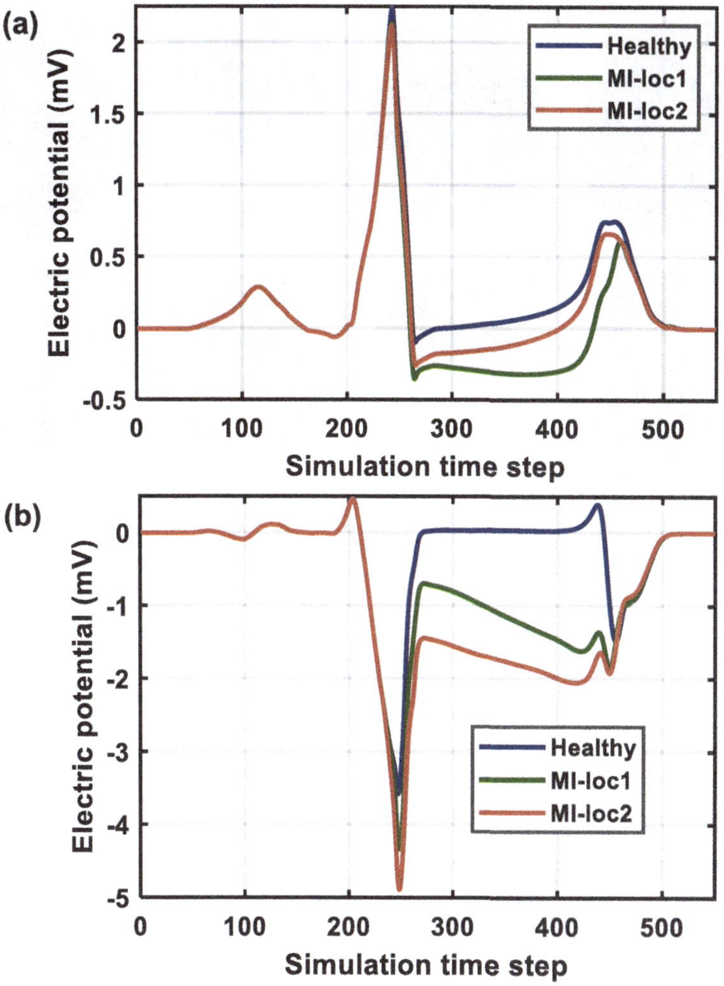

Fig. 2.14 (**a**) Lead-V_I and (**b**) Lead-V_2 ECG signals under healthy, MI-loc1, and MI-loc2 cardiac conditions

will emit electrical impulses more frequently than SAN. Figure 2.15 illustrates the excitation and propagation of cardiac electrodynamics under AF condition. The electrical waves generated from the AF trigger interact with that generated from SAN, resulting in uncoordinated cardiac electrodynamics (i.e., spiral wave as shown in Fig. 2.15d) and further leading to irregular depolarization and repolarization in both the atria and ventricles.

Figure 2.16a and b shows the simulated Lead-V_I and Lead-V_2 ECG waveforms under healthy and AF cardiac conditions, respectively. The ECG cycle in both leads is aperiodic and irregular in the AF heart compared with the healthy heart due to the fact that the AF trigger impacts the depolarization and repolarization

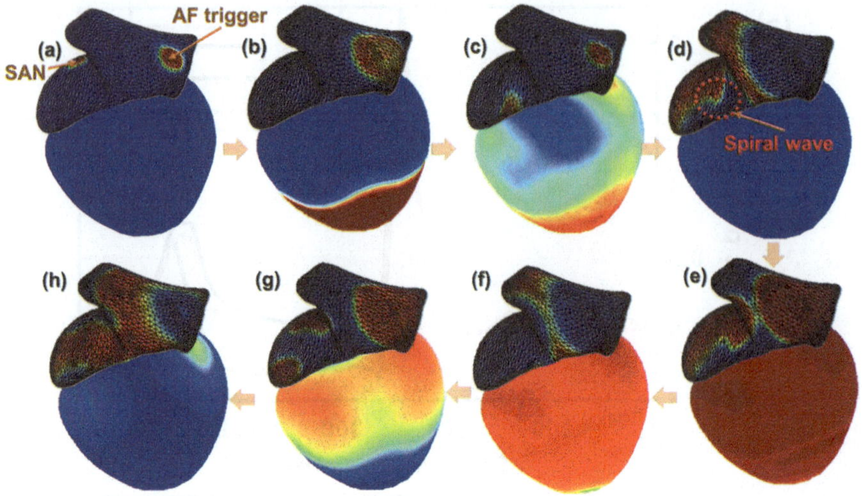

Fig. 2.15 The simulated excitation and propagation of cardiac electrodynamics with atrial fibrillation

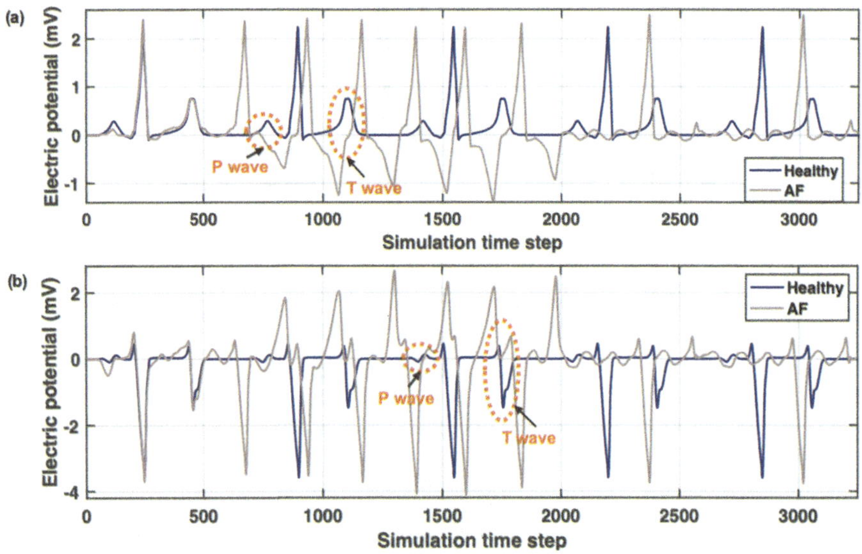

Fig. 2.16 (**a**) Lead-V_1 and (**b**) Lead-V_2 ECG signals with 5 cycles under healthy and AF cardiac conditions

of both the atria and ventricles. Specifically, the morphology of the P wave, QRS complex, and T wave in AF-altered ECG signal is significantly different in the amplitude, shape, and timing from that in a normal heart, which is consistent with the clinical findings for AF diagnosis [44]. Thus, the proposed simulation model has successfully replicated the AF-altered cardiac electrodynamics and ECG signal and can further be potentially implemented to quantitatively investigate the AF mechanism and design simulation–optimization algorithms for optimal surgical procedure (e.g., optimal ablation sequence) for AF patients.

Discussion: Parameter Setting in Cardiac Simulation

The parameter setting for the simulation with healthy heart in this chapter is based on the general assumptions on model parameters that are taken from the existing literature [21, 29]. However, cardiac simulation is subject to variabilities among different individuals even under healthy heart condition, which manifests in the variations of model parameters. Personalized cardiac simulation (i.e., true parameter setting for each individual) requires estimating electrophysiological parameters and reconstructing heart geometry from personalized medical data (e.g., patient-specific ECG signals, MRI images, CT-scan), which will be covered in the next subsection. Here, we further investigate how the variations in model parameters impact the simulated body-surface ECG signals. Specifically, we vary the value of each parameter in the A-P model (i.e., Eq. (1)–(3)) by 10% over its nominal setting provided in Table 2.1. Figure 2.17a shows the simulated Lead-V_I ECG signals given different parameter settings. Note that the simulation is conducted in a healthy heart (i.e., without heterogeneous pathological variations in the parameter setting), and the simulated ECG signals show a similar overall pattern. However, the magnitude and heartbeat length are different due to the variations in the model parameters. Additionally, we investigate the effect of the size of heart geometry on the simulated body-surface ECG signals. Figure 2.17b shows the variations in the Lead-V_I ECG with different heart sizes, i.e., original size, with 10% increase, and with 10% reduction. Table 2.2 compares the root-mean-square deviation (RMSD) between the signals generated from the nominal setting and the new parameter settings. Note that a 10% variation in parameter a or k incurs a larger deviation from the baseline signal compared to other settings. In other words, the simulation result of cardiac electrophysiological modeling is more sensitive to parameters a and k.

It is also worth noting that we assign a small conductivity (i.e., $D(AVN) = 0.5\,\mathrm{mS}\cdot\mathrm{m}^{-1}$) to AVN in order to guarantee that both atria are fully depolarized before the ventricles start to depolarize. We further investigate the impact of $D(AVN)$ on the simulated ECG signal as shown in Fig. 2.18. Note that when a bigger conductivity (i.e., a small delay in the electrodynamics between atria and ventricles) is assigned to AVN, the ventricles start to depolarize before the atria complete the depolarization as shown by the green curve. On the other hand, when AVN is assigned with a conductivity that is too small, an extra delay will be incurred and the ventricles start to depolarize after the atria complete both the depolarization and repolarization (as shown by the red curve), which is not consistent with the true electrophysiological activity of the heart [18].

In the simulation with MI, we set the conductivity in MI regions as zero (i.e., $D(MI) = 0$) to represent the non-excitability of the infarcted tissue. In fact, MI not only can happen at different locations in the heart but also can evolve dynamically over time. For example, the temporal progression of MI is through

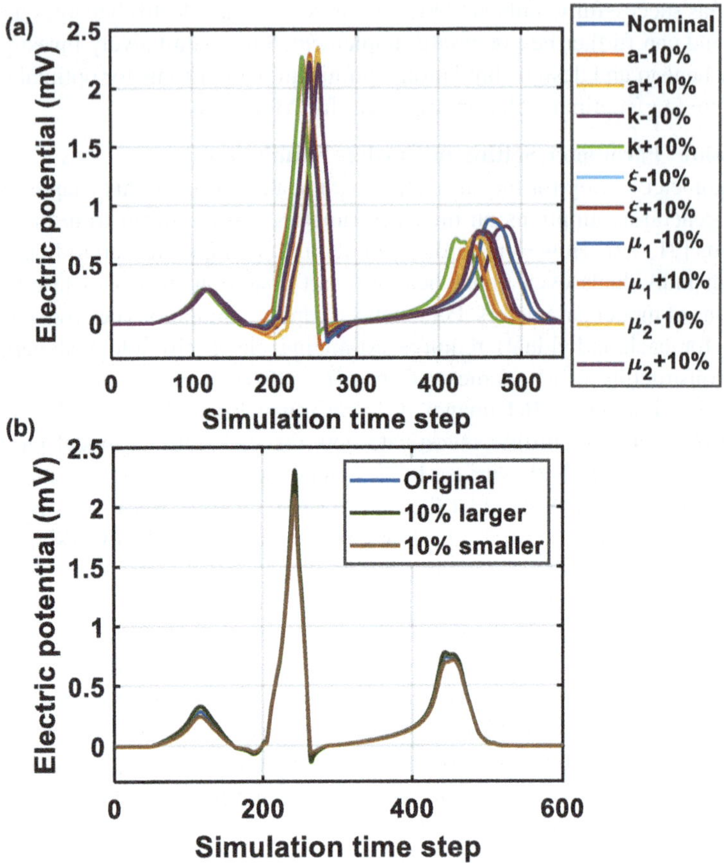

Fig. 2.17 The variations of simulated Lead-V_I ECG signal with respect to (**a**) the values of model parameters; (**b**) the size of heart geometry

Table 2.2 RMSD between Lead-V_I ECGs generated from the nominal setting and the new parameter settings

RMSD	a	k	ξ_0	μ_1	μ_2	Geometry Size
+10%	6.23	7.34	0.20	2.95	0.73	0.34
−10%	6.39	6.44	0.19	2.40	0.73	0.57

the process from ischemia (least severe), injury (moderately severe), to necrosis (most severe). The most severe MI, i.e., necrosis tissue, is non-excitable, which is set with a zero conductivity, i.e., $D(MI) = 0\,\text{mS} \cdot \text{m}^{-1}$. We further investigate the impact of MI severity on the simulated ECG signal as shown in Fig. 2.19. Note that the reduction of the cardiac tissue conductivity introduces pattern variations compared with the normal signal (i.e., $D = 8\,\text{mS} \cdot \text{m}^{-1}$), and the most severe MI (i.e., $D(MI) = 0\,\text{mS} \cdot \text{m}^{-1}$) incurs the largest deflection in the simulated ECG signal.

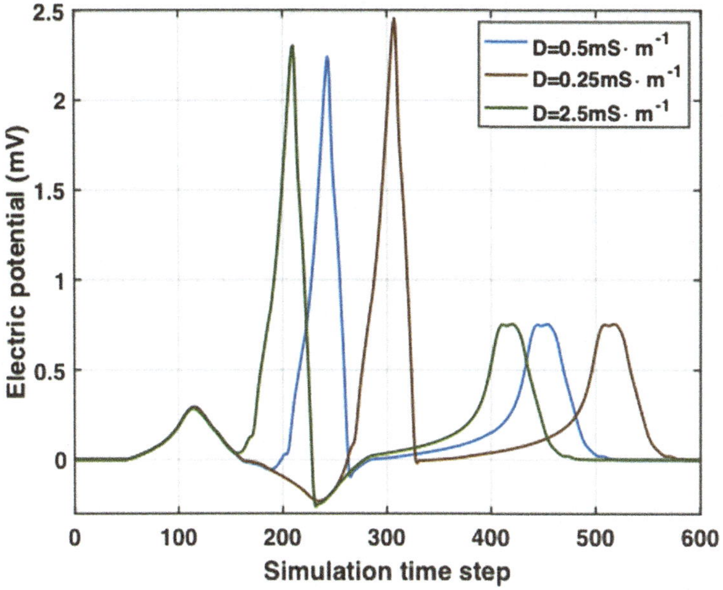

Fig. 2.18 The variations of simulated Lead-V_I ECG signal with respect to the conductivity value of AVN (i.e., $D(AVN)$)

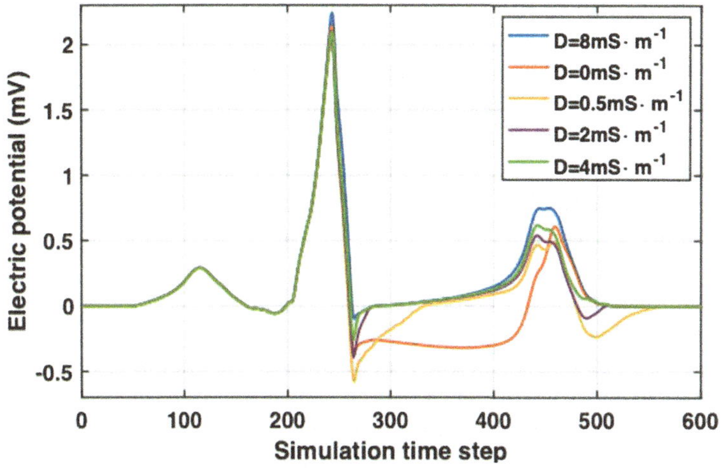

Fig. 2.19 The variations of simulated Lead-V_I ECG with respect to the severity degree of MI at MI-loc1

2.4 Calibration of 3D Cardiac Simulation

The physics-based cardiac model contains unknown parameters that need to be calibrated before making predictions. Model parameter values are generally taken from the literature on similar studies, which is incapable to accurately characterize the electrophysiological properties of different patients. It is critical to leverage medical sensing data to calibrate the cardiac model to make the simulation-based prediction close to reality. The widely used approach for computer model calibration is developed by Kennedy and O'Hagan [45], which has a variety of adapted formulations to solve different problems [46–48]. These approaches are primarily developed for models with univariate (scalar) output, which are difficult to be extended for model calibration with high-dimensional output. Cardiac electrical signals propagate dynamically over spatial and temporal domains, leading to high-dimensional space–time data. Our prior work [26] has developed a dynamic spatiotemporal warping (D-ST-W) method to quantify the spatiotemporal dissimilarity between the real cardiac system and the 3D simulation model and further proposed a hierarchical Gaussian process (HGP) approach to predict the D-ST-W discrepancy for effective simulation calibration.

Dynamic Spatiotemporal Warping (D-ST-W) for Discrepancy Measure
The simulation output can be written as

$$\boldsymbol{Y}_{\boldsymbol{\theta}}^{s} = \mathcal{F}^{\boldsymbol{\theta}}, \tag{2.21}$$

where $\mathcal{F}$ denotes the cardiac simulation model and $\boldsymbol{\theta}$ represents the model parameter that needs to be calibrated. We further denote the physical measurement as $\boldsymbol{Y}^{p}$. Both $\boldsymbol{Y}_{\boldsymbol{\theta}}^{s}$ and $\boldsymbol{Y}^{p}$ are space–time matrices describing the cardiac electrodynamics, i.e., $\boldsymbol{Y}_{\boldsymbol{\theta}}^{s} = \boldsymbol{Y}_{\boldsymbol{\theta}}^{s}(s,t)$ and $\boldsymbol{Y}^{p} = \boldsymbol{Y}^{p}(s,t)$, where s and t denote the spatial and temporal locations, respectively. The true model parameter, denoted as $\boldsymbol{\theta}^{*}$, is defined as the one generating a simulation response $\boldsymbol{Y}_{\boldsymbol{\theta}^{*}}^{s}$ close to $\boldsymbol{Y}^{p}$ up to a Gaussian error $\boldsymbol{\epsilon}$, i.e.,

$$\boldsymbol{Y}^{p}(s,t) = \boldsymbol{Y}_{\boldsymbol{\theta}^{*}}^{s}(s,t) + \boldsymbol{\epsilon}(s,t), \tag{2.22}$$

where $\boldsymbol{\epsilon}(s,t) \sim \mathcal{N}(0,\sigma^{2})$. In other words, model calibration aims to search for the optimal parameter set, $\boldsymbol{\theta}^{*}$, such that the output $\boldsymbol{Y}_{\boldsymbol{\theta}^{*}}^{s}$ from the simulator is most similar to $\boldsymbol{Y}^{p}$ from physical measurements. Due to the spatiotemporal structure, $\boldsymbol{Y}^{s}(s,t)$ and $\boldsymbol{Y}^{p}(s,t)$ may be misaligned with possible mismatches between the numerical solution domain and real space–time domain. This misalignment will result in a big Euclidean distance even if the two sets of signals are identical, leading to inaccurate model calibration.

Given two signals, $\boldsymbol{Y}^{s} \in \mathcal{R}^{S\times T^{s}}$ (note $\boldsymbol{\theta}$ is omitted hereafter for notation convenience) and $\boldsymbol{Y}^{p} \in \mathcal{R}^{S\times T^{p}}$, S is the cardinality of the spatial nodes, and T^{s} and T^{p} are the number of time stamps in the simulation outputs and sensor measurements, respectively, and the D-ST-W method spatiotemporally aligns $\boldsymbol{Y}^s$ and $\boldsymbol{Y}^p$ by minimizing the following cost function:

$$J_{\text{DSTW}}(\boldsymbol{W}^s, \boldsymbol{W}^p, \boldsymbol{q}, \boldsymbol{h}) = \|(\boldsymbol{Q}\boldsymbol{Y}^s + \boldsymbol{H}\mathbf{1}_{S\times T^s})\boldsymbol{W}^s - \boldsymbol{Y}^p\boldsymbol{W}^p\|_F^2 \\ +\lambda_q^2\|\Delta_s\boldsymbol{q}\|^2 + \lambda_h^2\|\Delta_s\boldsymbol{h}\|^2, \tag{2.23}$$

where $\boldsymbol{W}^s$ and $\boldsymbol{W}^p$ denote the temporal warping matrices; $\boldsymbol{Q}$ and $\boldsymbol{H}$ perform the spatial warping (both are diagonal matrices formed by the scale vector $\boldsymbol{q}$ and offset vector $\boldsymbol{h}$); $\mathbf{1}_{S\times T^s}$ is a matrix of ones; Δ_s is a surface Laplacian to regularize the spatial smoothness of $\boldsymbol{q}$ and $\boldsymbol{h}$; λ_q and λ_h are parameters to control the spatial regularization.

Minimizing the cost function in Eq. (2.23) will generate an optimal spatiotemporal alignment ($\boldsymbol{W}^{s*}$, $\boldsymbol{W}^{p*}$, $\boldsymbol{q}^*$, $\boldsymbol{h}^*$), based on which we derive the model discrepancy as

$$\delta(\boldsymbol{\theta}) = \left\|\left(\boldsymbol{Q}_{\boldsymbol{\theta}}^*\boldsymbol{Y}_{\boldsymbol{\theta}}^S + \boldsymbol{H}_{\boldsymbol{\theta}}^*\mathbf{1}_{S\times T^s}\right)\boldsymbol{W}_{\boldsymbol{\theta}}^{s*} - \boldsymbol{Y}^p\boldsymbol{W}_{\boldsymbol{\theta}}^{p*}\right\|_F. \tag{2.24}$$

Note that $\delta(\boldsymbol{\theta})$ will be small if $\boldsymbol{\theta}$ is close to the truth (i.e., $\boldsymbol{\theta}^*$) and vice versa. As such, we can estimate the optimal parameter as the one minimizing $\delta(\boldsymbol{\theta})$, i.e.,

$$\hat{\boldsymbol{\theta}} = \arg\min_{\boldsymbol{\theta}} \delta(\boldsymbol{\theta}). \tag{2.25}$$

The estimation accuracy of $\hat{\boldsymbol{\theta}}$ depends on the number of simulation runs given different settings of the model parameter. However, large-scale cardiac simulation is computationally demanding. It is infeasible to run exhaustive simulation trials for accurate parameter estimation. In our prior work, we proposed to construct a regression model $\hat{\delta}(\boldsymbol{\theta})$ and treat it as a surrogate for $\delta(\boldsymbol{\theta})$, and further estimate the model parameter by minimizing $\hat{\delta}(\boldsymbol{\theta})$.

Hierarchical Gaussian Process (HGP) Modeling of $\delta(\boldsymbol{\theta})$

Surrogate modeling based on GP has wide applications in simulation calibration [25, 49, 50]. However, conventional GPs typically assume the response surface exhibits global smoothness, which can blur critical elements of the response variable and is not effective to capture abrupt changes. Cardiac simulation is highly complex and nonlinear due to the intricate geometric structure and spatiotemporal dynamic process. Hence, a global GP often fails to adapt to varying levels of smoothness to characterize detailed features of the simulation response, leading to an inaccurate representation of the cardiac simulation model. Our prior work developed a HGP to model $\delta(\boldsymbol{\theta})$ to capture both the global and local structures of the simulation response.

The parameter setting for the cardiac simulator, i.e., $\boldsymbol{\theta}$'s with $\boldsymbol{\theta} \in \mathcal{X}$, serves as the input of HGP, and the corresponding output is the discrepancy measure, i.e., $\delta(\boldsymbol{\theta})$'s. The HGP model is defined as

$$\delta(\boldsymbol{\theta}) = \hat{\delta}(\boldsymbol{\theta}) + \epsilon_\delta = \hat{\delta}_{global}(\boldsymbol{\theta}) + \hat{\delta}_{local}(\boldsymbol{\theta}) + \epsilon_\delta, \quad \epsilon_\delta \sim \mathcal{N}\left(0, \sigma_\epsilon^2\right), \tag{2.26}$$

where $\hat{\delta}_{global}(\boldsymbol{\theta})$ is a global GP defined on the entire input space (i.e., $\mathcal{X}$) to capture the global smoothness of the response variable. $\hat{\delta}_{local}(\boldsymbol{\theta})$ is a local GP defined on the subspace of $\mathcal{X}$ to characterize local variabilities and improve the overall prediction performance. Given the collected observations $\{\boldsymbol{\Theta}_n, \boldsymbol{\delta}_n\}$ after n rums of computer simulation, where $\boldsymbol{\Theta}_n = [\boldsymbol{\theta}_1, \boldsymbol{\theta}_2, \ldots, \boldsymbol{\theta}_n]'$ and $\boldsymbol{\delta}_n = [\delta(\boldsymbol{\theta}_1), \delta(\boldsymbol{\theta}_2), \ldots, \delta(\boldsymbol{\theta}_n)]'$, the estimated probability distribution of $\delta(\boldsymbol{\theta})$ at an arbitrarily unsimulated parameter setting $\boldsymbol{\theta}$ is derived as

$$\delta(\boldsymbol{\theta})|\boldsymbol{\theta}, \boldsymbol{\Theta}_n, \boldsymbol{\delta}_n \sim \mathcal{N}\left(\mu_n(\boldsymbol{\theta}), \sigma_n^2(\boldsymbol{\theta})\right), \tag{2.27}$$

where $\mu_n(\boldsymbol{\theta})$ and $\sigma_n^2(\boldsymbol{\theta})$ are the predictive mean and variance, respectively.

The posterior mean (i.e., $\mu_n(\boldsymbol{\theta})$) will serve as the statistical surrogate such that the discrepancy metric at an arbitrary parameter setting can be estimated without actually running the simulator, which can be leveraged to intuitively estimate the model parameter as $\hat{\boldsymbol{\theta}} = \arg\min_{\boldsymbol{\theta}} \mu_n(\boldsymbol{\theta})$. However, the accuracy of HGP modeling depends on the quality and quantity of training points, which requires repeated evaluations of the cardiac simulator. Increasing the computational efficiency and prediction accuracy calls upon an efficient strategy to collect the most informative training points. To cope with this issue, we proposed an active learning method to construct the HGP surrogate with high fidelity. It includes two main steps: (1) Training points collection: The informative training points will help not only identify the small $\delta(\boldsymbol{\theta})$ region but also enable the HGP to approximate $\delta(\boldsymbol{\theta})$ globally well. Thus, such points need to balance the exploitation of current knowledge and the exploration of the uncertain space. This is accomplished by using the upper-confidence-bound (UCB) criterion [88], i.e.,

$$\boldsymbol{\theta}_{n+1} = \arg\max_{\boldsymbol{\theta}}\{-\mu_n(\boldsymbol{\theta}) + \sqrt{c \cdot n} \cdot \sigma_n(\boldsymbol{\theta})\}, \tag{2.28}$$

where c is a constant to balance the exploration and exploitation, and $\mu_n(\boldsymbol{\theta})$ and $\sigma_n(\boldsymbol{\theta})$ are the estimated mean and standard deviation based on the first n training points. (2) Update HGP: Once a new point $\boldsymbol{\theta}_{n+1}$ is selected, the cardiac simulation will be executed to calculate $\delta(\boldsymbol{\theta}_{n+1})$. Then, the HGP will be updated with the new training point $\{\boldsymbol{\theta}_{n+1}, \delta(\boldsymbol{\theta}_{n+1})\}$. These two steps will repeat iteratively until convergence or the computation resource is used up. The optimal model parameter will be finally estimated as

$$\hat{\boldsymbol{\theta}} = \arg\min_{\theta}\{\mu_\delta(\boldsymbol{\theta})\}, \tag{2.29}$$

where $\mu_\delta(\boldsymbol{\theta})$ stands for the posterior mean of the final HGP.

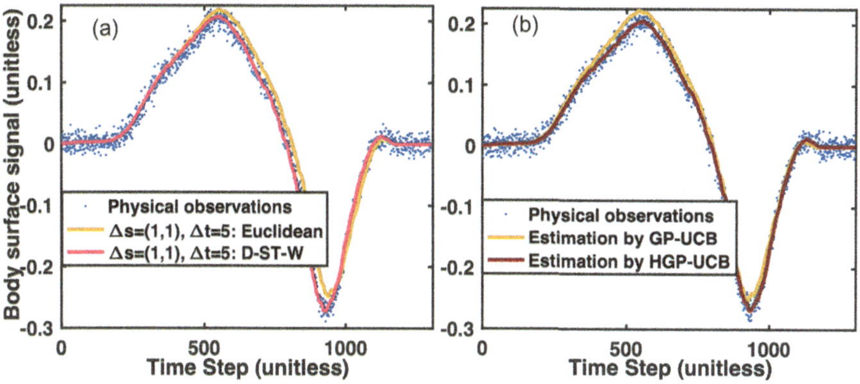

Fig. 2.20 (**a**) The comparison between physical observations and estimated signals (**a**) based on Euclidean dissimilarity and D-ST-W metric under the spatiotemporal misalignment $(\Delta s, \Delta t)$; (**b**) from GP-UCB and HGP-UCB when $\gamma = 0.1$ in the physical observation

Our prior work has evaluated the performance of the spatiotemporal warping enabled HGP-UCB (ST-HGP-UCB) calibration framework in the 3D cardiac simulation. The cardiac simulation is conducted on a 3D torso-heart geometry: the heart geometry is formed by 3179 nodes and 6343 mesh elements, and the torso geometry consists of 352 nodes and 677 mesh elements. To generate the physical measurements, a "true" setting of the parameter, i.e., $\boldsymbol{\theta}^*(s)$, is assigned to the cardiac simulator, and then the physical observation is obtained as $\boldsymbol{Y}^p(s,t) = \boldsymbol{Y}^s_{\theta^*}(s,t) + \boldsymbol{\epsilon}(s,t)$, where $\boldsymbol{\epsilon}(s,t) \sim \mathcal{N}(0, \gamma^2 \cdot \sigma^2_{\boldsymbol{Y}^s_{\theta^*}})$, γ denotes the noise level, $\sigma^2_{\boldsymbol{Y}^s_{\theta^*}}$ is the variance of $\boldsymbol{Y}^s_{\theta^*}$, and $\boldsymbol{Y}^s_{\theta^*}$ denotes the simulated BSPMs. The ST-HGP-UCB method is then implemented to estimate $\boldsymbol{\theta}$ such that the discrepancy between $\boldsymbol{Y}^s_{\theta}(s,t)$ and $\boldsymbol{Y}^p(s,t)$ is minimized.

Figure 2.20a compares the performance in model calibration between HGP-UCB based on Euclidean-dissimilarity measure and HGP-UCB based on D-ST-W metric under different levels of space–time misalignment with a noise level of $\gamma = 0.1$. Note that the HGP-UCB based on D-ST-W yields the estimated signals closer to the physical measurements compared with the approach based on Euclidean dissimilarity. Hence, the D-ST-W metric effectively alleviates the impact of spatiotemporal misalignments between the simulation output and the physical observation on the accuracy of simulation calibration. Figure 2.20b further illustrates the comparisons of calibration performance between GP-UCB and HGP-UCB algorithms when $\gamma = 0.1$. Note that the estimated signals from the HGP-UCB methods are more similar to the physical measurements, demonstrating superior performance in calibration 3D cardiac simulation compared with commonly used GP-UCB approach.

References

1. Y. Rudy, J.R. Silva, Computational biology in the study of cardiac ion channels and cell electrophysiology. Q. Rev. Biophys. **39**(1), 57–116 (2006)
2. V.E. Bondarenko, G.P. Szigeti, G.C. Bett, S.-J. Kim, R.L. Rasmusson, Computer model of action potential of mouse ventricular myocytes. Am. J. Physiol. Heart Circ. Physiol. **287**(3), H1378–H1403 (2004)
3. D. Du, H. Yang, S.A. Norring, E.S. Bennett, In-silico modeling of glycosylation modulation dynamics in hERG ion channels and cardiac electrical signals. IEEE J. Biomed. Health Inform. **18**(1), 205–214 (2013)
4. K.H. Ten Tusscher, A.V. Panfilov, Cell model for efficient simulation of wave propagation in human ventricular tissue under normal and pathological conditions. Phys. Med. Biol. **51**(23), 6141 (2006)
5. R. Clayton, O. Bernus, E. Cherry, et al., Models of cardiac tissue electrophysiology: Progress, challenges and open questions. Prog. Biophys. Mol. Biol. **104**(1–3), 22–48 (2011)
6. D. Noble, Modeling the heart—from genes to cells to the whole organ. Science **295**(5560), 1678–1682 (2002)
7. E.J. Vigmond, F. Aguel, N.A. Trayanova, Computational techniques for solving the bidomain equations in three dimensions. IEEE Trans. Biomed. Eng. **49**(11), 1260–1269 (2002)
8. D. Du, H. Yang, S.A. Norring, E.S. Bennett, Multi-scale modeling of glycosylation modulation dynamics in cardiac electrical signaling, in *2011 Annual International Conference of the IEEE Engineering in Medicine and Biology Society* (IEEE, New York, 2011), pp. 104–107
9. S.A. Norring, A.R. Ednie, T.A. Schwetz, D. Du, H. Yang, E.S. Bennett, Channel sialic acids limit hERG channel activity during the ventricular action potential. FASEB J. **27**(2), 622–631 (2013)
10. M.L. Montpetit, P.J. Stocker, T.A. Schwetz, et al., Regulated and aberrant glycosylation modulate cardiac electrical signaling. Proc. Natl. Acad. Sci. **106**(38), 16517–16522 (2009)
11. D. Du, H. Yang, A.R. Ednie, E.S. Bennett, In-silico modeling of the functional role of reduced sialylation in sodium and potassium channel gating of mouse ventricular myocytes. IEEE J. Biomed. Health Inform. **22**(2), 631–639 (2017)
12. H. Kim, H. Yang, A.R. Ednie, E.S. Bennett, Simulation modeling of reduced glycosylation effects on potassium channels of mouse cardiomyocytes, in *Frontiers in Physiology* (2022), p. 272
13. K.H. Ten Tusscher, D. Noble, P.-J. Noble, A.V. Panfilov, A model for human ventricular tissue, in *American Journal of Physiology-Heart and Circulatory Physiology* (2004)
14. K.H. Ten Tusscher, A.V. Panfilov, Alternans and spiral breakup in a human ventricular tissue model. Am. J. Physiol. Heart Circ. Physiol. **291**(3), H1088–H1100 (2006)
15. D. Du, H. Yang, A.R. Ednie, E.S. Bennett, Statistical metamodeling and sequential design of computer experiments to model glyco-altered gating of sodium channels in cardiac myocytes. IEEE J. Biomed. Health Inform. **20**(5), 1439–1452 (2015)
16. M.A. Quiroz-Juarez, O. Jimenez-Ramirez, R. Vazquez-Medina, E. Ryzhii, M. Ryzhii, J.L. Aragon, Cardiac conduction model for generating 12 lead ECG signals with realistic heart rate dynamics. IEEE Trans. NanoBioscience **17**(4), 525–532 (2018)
17. M. Quiroz-Juarez, O. Jimenez-Ramirez, R. Vazquez-Medina, V. Brena-Medina, J. Aragon, R. Barrio, Generation of ECG signals from a reaction-diffusion model spatially discretized. Sci. Rep. **9**(1), 1–10 (2019)
18. G.J. Tortora, B.H. Derrickson, *Principles of Anatomy and Physiology* (Wiley, New York, 2018)
19. M. Balakrishnan, V.S. Chakravarthy, S. Guhathakurta, Simulation of cardiac arrhythmias using a 2D heterogeneous whole heart model. Front. Physiol. **6**, 374 (2015)
20. E.M. Izhikevich, R. FitzHugh, FitzHugh-Nagumo model. Scholarpedia **1**(9), 1349 (2006)
21. S. Sovilj, R. Magjarević, N.H. Lovell, S. Dokos, A simplified 3D model of whole heart electrical activity and 12-lead ECG generation, in *Computational and Mathematical Methods in Medicine*, vol. 2013 (2013)

22. E. Schenone, A. Collin, J.-F. Gerbeau, Numerical simulation of electrocardiograms for full cardiac cycles in healthy and pathological conditions. Int. J. Numer. Methods Biomed. Eng. **32**(5), e02744 (2016)
23. R.R. Aliev, A.V. Panfilov, A simple two-variable model of cardiac excitation. Chaos, Solitons & Fractals **7**(3), 293–301 (1996)
24. M.P. Nash, A.V. Panfilov, Electromechanical model of excitable tissue to study reentrant cardiac arrhythmias. Prog. Biophys. Mol. Biol. **85**(2–3), 501–522 (2004)
25. J. Dhamala, H.J. Arevalo, J. Sapp, et al., Spatially adaptive multi-scale optimization for local parameter estimation in cardiac electrophysiology. IEEE Trans. Med. Imaging **36**(9), 1966–1978 (2017)
26. B. Yao, Spatiotemporal modeling and optimization for personalized cardiac simulation. IISE Trans. Healthcare Syst. Eng. **11**(2), 1–35 (2021)
27. M.J. Ackerman, The visible human project. Proc. IEEE **86**(3), 504–511 (1998)
28. A. Fedorov, R. Beichel, J. Kalpathy-Cramer, et al., 3D slicer as an image computing platform for the quantitative imaging network. Magn. Reson. Imaging **30**(9), 1323–1341 (2012)
29. J. Malmivuo, R. Plonsey, et al., *Bioelectromagnetism: Principles and Applications of Bioelectric and Biomagnetic Fields* (Oxford University Press, USA, 1995)
30. J. Corral-Acero, F. Margara, M. Marciniak, et al., The 'Digital Twin' to enable the vision of precision cardiology. Eur. Heart J. **41**(48), 4556–4564 (2020)
31. G. Liu, J.B. Iden, K. Kovithavongs, R. Gulamhusein, H.J. Duff, K.M. Kavanagh, In vivo temporal and spatial distribution of depolarization and repolarization and the illusive murine T wave. J. Physiol. **555**(1), 267–279 (2004)
32. P.C. Franzone, L. Pavarino, S. Scacchi, B. Taccardi, Modeling ventricular repolarization: Effects of transmural and apex-to-base heterogeneities in action potential durations. Math. Biosci. **214**(1–2), 140–152 (2008)
33. E.W. Dijkstra et al., A note on two problems in connexion with graphs. Numer. Math. **1**(1), 269–271 (1959)
34. D.E. Hurtado, E. Kuhl, Computational modelling of electrocardiograms: Repolarisation and t-wave polarity in the human heart. Comput. Methods Biomech. Biomed. Engin. **17**(9), 986–996 (2014)
35. Z. Qu, A. Garfinkel, An advanced algorithm for solving partial differential equation in cardiac conduction. IEEE Trans. Biomed. Eng. **46**(9), 1166–1168 (1999)
36. B. Yao, H. Yang, Physics-driven spatiotemporal regularization for high-dimensional predictive modeling: A novel approach to solve the inverse ECG problem. Sci. Rep. **6**, 39012 (2016)
37. B. Yao, S. Pei, H. Yang, Mesh resolution impacts the accuracy of inverse and forward ECG problems, in *2016 38th Annual International Conference of the IEEE Engineering in Medicine and Biology Society (EMBC)* (IEEE, New York, 2016), pp. 4047–4050
38. R.H. Bartels, G.H. Golub, The simplex method of linear programming using LU decomposition. Commun. ACM **12**(5), 266–268 (1969)
39. F. Holmqvist, J. Carlson, P.G. Platonov, Detailed ECG analysis of atrial repolarization in humans. Ann. Noninvasive Electrocardiol. **14**(1), 13–18 (2009)
40. B. Yao, R. Zhu, H. Yang, Characterizing the location and extent of myocardial infarctions with inverse ECG modeling and spatiotemporal regularization. IEEE J. Biomed. Health Inform. **22**(5), 1445–1455 (2017)
41. K.F. Decker, Y. Rudy, Ionic mechanisms of electrophysiological heterogeneity and conduction block in the infarct border zone. Am. J. Physiol. Heart Circ. Physiol. **299**(5), H1588–H1597 (2010)
42. R.S. Gibson, R. Crampton, D.D. Watson, et al., Precordial ST-segment depression during acute inferior myocardial infarction: clinical, scintigraphic and angiographic correlations. Circulation **66**(4), 732–741 (1982)
43. M. Haissaguerre, P. Jais, D.C. Shah, et al., Spontaneous initiation of atrial fibrillation by ectopic beats originating in the pulmonary veins. N. Engl. J. Med. **339**(10), 659–666 (1998)

44. P. Kirchhof, S. Benussi, D. Kotecha, et al., 2016 ESC guidelines for the management of atrial fibrillation developed in collaboration with EACTS. European Journal of Cardiothoracic Surgery **50**(5), e1–e88 (2016)
45. M.C. Kennedy, A. O'Hagan, Bayesian calibration of computer models. J. R. Stat. Soc. Ser. B (Stat Methodol.) **63**(3), 425–464 (2001)
46. M. Goldstein, J. Rougier, Bayes linear calibrated prediction for complex systems. J. Am. Stat. Assoc. **101**(475), 1132–1143 (2006)
47. V.R. Joseph, S.N. Melkote, Statistical adjustments to engineering models. J. Qual. Technol. **41**(4), 362–375 (2009)
48. D. Higdon, M. Kennedy, J.C. Cavendish, J.A. Cafeo, R.D. Ryne, Combining field data and computer simulations for calibration and prediction. SIAM J. Sci. Comput. **26**(2), 448–466 (2004)
49. S. Ba, V.R. Joseph, et al., Composite Gaussian process models for emulating expensive functions. Ann. Appl. Stat. **6**(4), 1838–1860 (2012)
50. B. Farmanesh, A. Pourhabib, B. Balasundaram, A. Buchanan, A Bayesian framework for functional calibration of expensive computational models through non-isometric matching. IISE Transactions **53**(3), 352–364, 352–364 (2020)

Chapter 3
Sensor-Based Modeling and Analysis of Cardiac Systems

Cardiac disorders happen every day and account for some 30% of mortalities in the USA. The sensor-based research for health informatics is targeting the timeliness of cardiovascular diagnostics or prognostics but is highly dependent on the close integration of computing, sensing, modeling methods with physiological processes to achieve medical systems with high levels of functionality, adaptability, autonomy, and effectiveness. This chapter presents the methodology of sensor-based modeling and analysis in guiding the optimal management of heart health. To that end, people's living styles and habits can be associated with sensor-based health variables (or biomarkers), providing education on heart-healthy living and raising the awareness of smart health. In general, these methods and tools do not depend on physics-based cardiac modeling and/or restrictive assumptions of system behavior and can thus be extended to the general population with heart diseases as well as other complex clinical conditions.

3.1 Electrocardiogram (ECG) Sensing

With rapid advancement of sensing and computing technology, sensor-based modeling methodology has evolved significantly for improving the quality of heart monitoring. In the clinical practice, there are often two broad categories of tests used to sense, monitor, and detect the early signs of cardiac disorders. Some are "static" measurements, while others are "dynamic" sensing:

- **Static measurement:**
  Static diagnostic tests are essentially frozen screenshots of cardiac functions or cardiac tissues, e.g., magnetic resonance imaging, chest X-ray, cardiac catheterization, and blood enzyme test. The static approaches involve very expensive tests and are not always readily available, especially in rural hospitals. Even if

H. Yang, B. Yao, *Sensing, Modeling and Optimization of Cardiac Systems*, SpringerBriefs in Service Science, https://doi.org/10.1007/978-3-031-35952-1_3

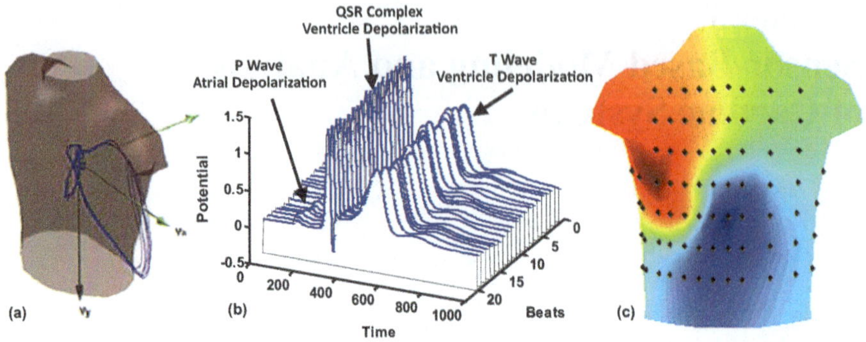

Fig. 3.1 (**a**) 3-lead vectorcardiogram (VCG) signals and the body torso model; (**b**) ECG signals and P, QRS, and T waves; (**c**) ECG imaging, also called body-surface potential mapping (BSPM)

the routine clinical examinations are performed multiple times per day, these intermittent measurements often fail to detect the life-threatening cardiac events.

- **Dynamic sensing:**
  Dynamic tests can obtain a wealth of information pertinent to heart dynamics continuously using physiological sensing and measurement devices, such as electrocardiogram (ECG) and electrocardiogram imaging (ECGI), as shown in Fig. 3.1. There is an increasing awareness that real-time operational details of cardiac functioning need to be tracked, as opposed to static measurements that are screenshots in cardiac processes. However, ECG interpretation is mainly conducted by healthcare professionals today as it was 50 years ago. The performance of human experts depends highly on the training, experience, and oftentimes on the memorization of ECG patterns for a variety of cardiac disorders. Also, experienced human experts are not always readily available, especially in rural or underdeveloped areas. But real-time ECG monitoring leads to a large amount of data in the modern healthcare environment. This makes it difficult for human experts to visually inspect large swathes of ECG signals for warning signs. These limitations have caused not just healthcare disparity issues but also the issues pertinent to the accuracy and timeliness of cardiovascular diagnostics, which may delay effective medical treatment.

As shown in Fig. 3.2, there are three broad classes of representation approaches: (a) temporal, (b) spatially varying temporal, and (c) spatiotemporal. The ECG signal in the form of temporal representation captures sequential cardiac electrical activity (also see Fig. 3.1b), which is initiated at the sinoatrial (SA) node to then excite the atrial muscle contraction, relayed in atrioventricular (AV) node to further propagate through bundle of His and Purkinje fibers toward ventricles depolarization and repolarization [1]. Willem Einthoven first assigned the letters P, Q, R, S, and T to various temporal wave deflections (see Fig. 3.1b). Traditional methods characterize the abnormal ECG patterns, e.g., the variations of P, QRS, and T waveforms for quantitative connections with cardiac disorders [2]. Additionally,

Methodology	Background
Temporal Representation	ECG characteristics Heart rate, P wave, QRS complex, ST-T wave, U wave, PR interval, QRS duration, QT interval, ST segment, etc.
Spatially-varying Temporal Representation	12 Lead ECG Bipolar limb leads: Lead I: Right Left, or Lateral Lead II: Superior Inferior Lead III: Superior Inferior Unipolar chest leads: Leads V1, V2, V3: Posterior Anterior Leads V4, V5, V6: Right Left, or Lateral Augmented unipolar limb leads: Lead aVR: Rightward Lead aVL: Leftward Lead aVF: Inferior
	3-lead VCG Right ↔ Left Superior ↔ Inferior Anterior ↔ Posterior
Spatiotemporal Representation	Noninvasive dynamic ECG imaging Spatiotemporal VCG Signal Representation

Fig. 3.2 ECG signal representation approaches

some sophisticated methods are used to transform the ECG signals in time [3], frequency [4, 5], time–frequency [6, 7], or state space domain [8–10] for the extraction of salient features sensitive to cardiac disorders.

However, one-lead ECG signals are only temporal projection of spatiotemporal cardiac electrical dynamics from one perspective or dimension. Thus, multiple-lead ECG systems (e.g., 3-lead vectorcardiogram (VCG) and/or 12-lead ECG) are designed to capture temporal projection of spatiotemporal cardiac electrical dynamics from multiple perspectives or dimensions. This, in turn, provides additional spatial information at a limited number of sites (see Fig. 3.1a). Figure 3.2 provides the detailed information pertinent to the representation of 12-lead ECG, 3-lead VCG, and ECG imaging (ECGI) data (see Fig. 3.1c). Because the heart is a 3D object and its electrical activity propagates and conducts across both space and time, let us denote the spatiotemporal electrical signals as $V(s,t) : s \in R \subset \mathbb{R}^d, t \in T$, where the dependence of spatial domain R on time T symbolizes the condition where electrical dynamics change in the spatial domain as well as over time. The state-of-the-art in spatiotemporal cardiac signal representation includes noninvasive electrocardiographic image [11] and dynamic spatiotemporal VCG signal representation [10]. Traditional reductionist's approaches for ECG analysis may be categorized as follows: (i) spatially varying time series model $V(s,t) = V_s(t)$, which visualizes and analyzes the temporal signals at each site or location. For example, clinicians examine the irregularity and abnormality in time-domain patterns of 12-lead ECG and 3-lead VCG; (ii) temporally varying spatial model $V(s,t) = V_t(s)$, which performs spatial analysis for all sensing sites at a specific time point, e.g., a snapshot of ECG image during the ventricular repolarization.

Undoubtedly, these approaches are following the reductionist's principles and focus on either the space given time or time given space (i.e., the so-called conditional methods). Nonetheless, space–time correlations are overlooked by the lenses of reductionists.

Due to the proliferation of multi-lead ECG signals that are varying both spatially and temporally, spatiotemporal modeling and analysis are urgently needed to uncover useful information that cannot be otherwise captured by conditional methods. Indeed, 1-D temporal projection of spatiotemporal signals diminishes important spatial information pertinent to cardiac pathological behaviors. As a result, such an information loss can negatively impact medical decisions that are made upon, especially when there is an need to detect early signs of cardiac ailments in its initial stage [10]. Little has been done to develop new spatiotemporal algorithms that identify the space–time disease processes, (i.e., spatial localization and temporal deterioration of myocardial infarctions). Although traditional statistics (i.e., non-spatial) can be applied to spatiotemporal cardiac signals, they may bring misleading conclusions due to the lack of spatial information gleaned from ECG sensors. This chapter focuses on the design of ECG biomarker algorithms that combine spatiotemporal statistics with increasingly available space–time data (i.e., time-varying ECG images from the network of distributed sensors) to predict the incidence of acute cardiac events and advance the early diagnostic capability of wearable ECG medical devices.

3.2 Modeling Incomplete and Uncertain Data

3.2.1 Introduction

Cardiovascular systems involve a great level of complexity and so are dynamic behaviors exhibited from cardiac functions. Therefore, modern healthcare facilities are investing in new generation of wireless and miniature sensors, data acquisition, and computing systems for advanced health informatics. ECG signals contain rich signatures and "memories" of cardiac electrical activity under a variety of conditions. Context-aware sensing provides the evolving dynamics of ambient parameters around the patient (e.g., blood pressure (BP), weight, blood sugar levels, active minutes, and dietary habits). As shown in Fig. 3.3, heterogeneous modalities of medical sensing bring a new form of order-3 tensor data with unique properties (i.e., variable heterogeneity, patient heterogeneity, and time asynchronization) [12, 13]. This is vastly different from the table form of data (e.g., consisting of predictor and response variables) that are commonly used in predictive modeling:

- Variable heterogeneity:
  It is common that different types of sensors are used to collaboratively capture a complete picture of the patient recovery process. For example, collaborative sensing involves various types of sensing variables (e.g., ECG, sleep pattern,

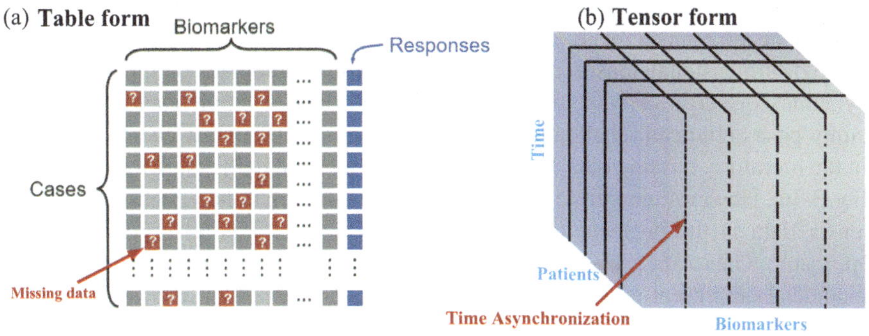

Fig. 3.3 (**a**) Traditional table form of data (e.g., biomarkers as predictors, outcomes as response variables) with the presence of missing data for predictive modeling, (**b**) New order-3 tensor-form data generated by heterogeneous modalities of medical sensing

medicine usage, active minutes, calories burned). Some are quantitative (continuous, discrete), while others may be qualitative (categorical, dummy variables).

- Patient heterogeneity:
  There are also heterogeneous types of patient populations, which may be classified by ages, gender, or procedure types. This provides an opportunity to investigate the disparity of recovery rate and quality of life for different patient populations.
- Time asynchronization:
  Data collection is patient-centered and often cannot be synchronized. For example, the frequency of measurements may be subjective to the information need of physicians or the patient's discretion. Although there are time stamps associated with collected data, these stamps are often not uniformly distributed along the time axis. Some variables may be recorded in an extremely low sampling rate, while others may be in a high sampling rate.

In addition, there are several uncertainty factors (e.g., human errors, noises, and artifacts) associated with ECG and context-aware sensing. Uncertainty factors deteriorate the quality of data recordings and the estimation performance of disease states. Human errors may be induced because of physical condition, attitude, emotions, and cognitive biases. Noises are random, uncontrollable, and unwanted disturbances that contaminate the useful information in the signals. Noises may be generated from the surrounding environment, sensors, or mobile devices. Artifacts refer to the noises within the frequency band of interest and exhibit similar morphologies as the signal itself, including ECG baseline wander, power-line interference, and electrode pop or contact noises. See more details about our prior works pertinent to the quality control of ECG signals and the interference mitigation of uncertainty factors in [14, 15]. In summary, missing and incomplete data are common in patient-centered sensing and measurement. This section presents new analytical methods and tools that help handle the heterogeneity in tensor-form data and further extract a parsimonious set of biomarkers for sensor-based optimization of heart health management.

High-dimensional tensor data provide rich information about the underlying dynamics of cardiovascular processes. However, data incompleteness and uncertainty pose significant challenges for information retrieval and decision modeling. In the literature, missing data imputation is often referred to the table form data (see Fig. 3.3a). However, very little has been done to model and impute high-dimensional tensor data in the presence of missing data (see Fig. 3.3b). As a nonparametric approach, Gaussian process (GP) is more flexible than traditional imputation methods to represent multi-dimensional covariance structure of time, variable, and patient. This section presents a review of GP models for missing data imputation in the context of high-dimensional tensors, thereby handling incompleteness and uncertainty in patient-centered sensing.

3.2.2 *Modeling Approaches*

Missing data are a general problem faced in healthcare analytics. This is mainly due to the fact that the collection of clinical data is often observational and subject to human discretion, as opposed to controlled experiments. In the literature, missing data imputation refers to a suite of analytical methods that handle the missingness problem so that the dataset can be further analyzed in the traditional way. The general idea is to predict the missing data by fully exploiting correlation structures hidden in the data. Examples of traditional imputation techniques include case deletion, hot deck imputation [16], K-nearest neighbors, functional regression [17], and machine learning methods such as support vector machines and deep learning [18]. However, these existing methods are more concerned about the imputation of table form data (e.g., biomarkers as predictors, outcomes as response variables) that are commonly used for predictive modeling (see Fig. 3.3a), and are less concerned about high-dimensional tensors (see Fig. 3.3b) generated by heterogeneous modalities of medical sensing.

Traditional table form data are static and not time-varying. In other words, temporal dimension is not specifically considered in most of existing methods for missing data imputation. As such, they are not well suited for the handling of order-3 tensor data that involve multi-dimensional covariance structure of time, variable, and patient. On the other hand, Kriging and GP models are commonly used in the domain of computer experiments and spatial analysis [19, 20]. An attractive feature is that both models provide the best linear unbiased prediction of missing values in a design space or a spatial region. Nonetheless, traditional Kriging and GP models assume the stationarity (i.e., constancy of mean, variance) in the underlying stochastic process. The stationary assumption limits the ability to deal with the order-3 tensor form of sensing variables that exhibit nonlinearity and non-stationarity. Also, multi-dimensional covariance structure significantly challenges the existing formulation of Kriging and GP models.

Our prior work developed a GP-recurrence model to predict the evolving dynamics of multivariate nonlinear time series [21]. Also, we coupled GP metamodels with incomplete data from biomedical experiments to predict complex response surface of large-scale simulation models of cardiac systems [22]. GP metamodels provide the uncertainty quantification of imputed predictions, which facilitate the identification of the next best design point for experiments. GP metamodeling minimizes the number of expensive experiments, while maximizing the prediction accuracy. In addition, our prior work developed an ICU postsurgical decision support system [12, 13]. ICU data are in the order-3 tensor form and involve technical challenges such as variable heterogeneity, patient heterogeneity, and time asynchronization. We leveraged GP models for missing data imputation. The GP model, defined as a collection of random variables, is treated as functional a priori on the time-varying ICU variables. In order to obtain the posterior distribution, this joint prior distribution is restricted to include only those functions that agree with incomplete ICU data. Model evaluation using data from 4000 ICU subjects showed that our proposed methods yield superior results over traditional approaches (i.e., Acute Physiology and Chronic Health Evaluation (APACHE), Sequential Organ Failure Assessment (SOFA) and Simplified Acute Physiology Score (SAPS)) [12, 13].

Furthermore, our prior work developed a nested Gaussian process to model tensor data [23, 24]. Note that there are 3 different dimensions, namely time, variables, and patients, in the tensor data (also see Fig. 3.3b). The new idea is to design a hierarchical model of 3-dimensional covariance structures in the nested Gaussian process. As shown in Fig. 3.3b, let us denote V variables are recorded and collected for P patients during the time period of T in the tensor data. Then $X_t(v_i, p_j)$ will be data value for the variable v_i from the patient p_j at the time point t.

First, the level-I GP model is constructed as $X_t(v_i, p_j) \sim GP(\mu_t, \mathbf{K}_t)$, where μ_t is the mean function and $\mathbf{K}_t$ is the covariance function (see Fig. 3.4). When data values are closer to each other over time (i.e., closeness in time), it is common that they tend to have a stronger correlation. Therefore, the covariance function $\mathbf{K}_t$ is designed as: $cov(X_{t_m}, X_{t_n})|_{v_i,p_j} = \sigma_t^2 \exp\left[\frac{-(t_m-t_n)^2}{2l_t^2(v_i,p_j)}\right] + \sigma_{nt}^2\delta(t_m, t_n)$, where σ_t^2 and σ_{nt}^2 are signal and noise variances in the dimension of time and $l_t(v_i, p_j)$ is the length scale.

Second, we will model μ_t using a level-II GP model as $\mu_t \sim GP(\mu_p, \mathbf{K}_p)$, where $\mathbf{K}_p$ represents the covariance between patients (e.g., p_m and p_n). This proposed hierarchical design handles nonstationarity in the underlying stochastic process through GP modeling of mean functions. In real-world environment, the joint distribution of tensor data will change over time. Traditional models with stationary assumptions (e.g., principal component analysis and regression models) are limited in their ability to readily address time-varying structures. The proposed level-II GP model leverages covariance structure among patients to model the non-stationary mean function μ_t. If two patients share closer values in clinical variables,

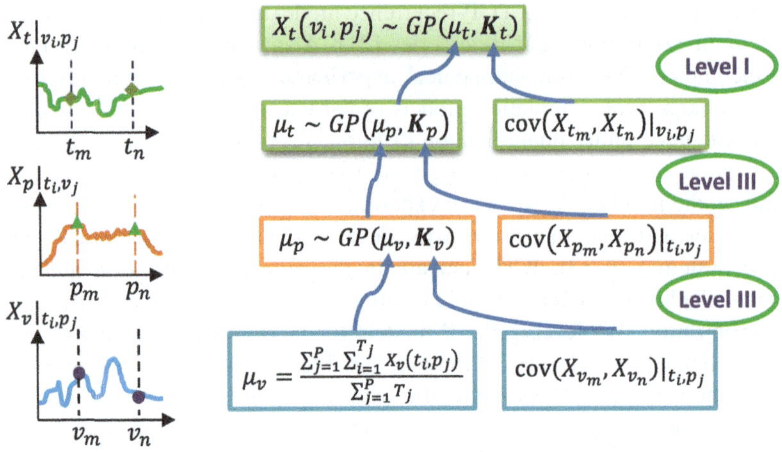

Fig. 3.4 Three-level hierarchical Gaussian process model for high-dimensional data modeling and imputation [23, 24]

they tend to have a stronger correlation. Therefore, the covariance function $\mathbf{K}_p$ is defined as: $cov(X_{p_m}, X_{p_n})|_{t_i, v_j} = \sigma_p^2 \exp\left[\frac{-(X_m - X_n)^2}{2l_p^2}\right] + \sigma_{np}^2 \delta(p_m, p_n)$.

Third, the level-III GP is constructed to model the mean function of μ_p as $\mu_p \sim GP(\mu_v, \mathbf{K}_v)$, where the covariance function $\mathbf{K}_v$ is to capture the similarity between clinical variables (e.g., v_m and v_n), and it is akin to the definition of $\mathbf{K}_p$. Here, μ_v is the average of tensor data across the time and patient dimensions, $\mu_v = \frac{\sum_{j=1}^{P} \sum_{i=1}^{T_j} X_v(t_i, p_j)}{\sum_{j=1}^{P} T_j}$ for a specific clinical variable. Here, T_j is the time duration for patient p_j. As opposed to 1-dimensional correlation in traditional models, the NGP model formulation captures 3-dimensional correlation structure (i.e., time, patients, and clinical variables) specific to order-3 tensor data. As a result, the NGP model offers a higher level of capability and flexibility to impute missing values in high-dimensional tensor data.

3.2.3 *Summary*

The high-dimensional imputation approach helps handle the issues of data incompleteness and uncertainty in the medical sensing environment. Specifically, the advantageous features include: (1) imputation modeling of order-3 tensor data (rather than table form data); (2) representation of nonstationary (or time-varying) mean functions (rather than stationary assumptions); (3) characterization of 3-dimensional correlation structure (i.e., time, patients, and clinical variables) specific to order-3 tensor data (rather than 1-dimensional correlation). This approach is not limited to ECG data, but rather can be generally applicable to high-dimensional tensor data analysis in a variety of other healthcare and engineering investigations.

3.3 Computationally Identify Sensory Biomarkers

3.3.1 Introduction

ECG signals cannot be directly used as predictors and have a much higher sampling rate than traditional clinical variables such as blood pressures, glucose levels, platelets, white blood cell count, and troponin. Biomarkers rather underlie the variations of ECG functional form. Although ECG is widely used in clinical practice, limited work has been done to identify biomarkers to track spatiotemporal variations of cardiovascular processes. Further, heterogeneous modalities of medical sensing bring the tensor form of data (see Fig. 3.3). High-dimensional tensor data pose significant challenges for the following tasks of statistical monitoring, risk prognostics, and decision optimization. This section presents new analytical methods and algorithms to extract ECG biomarkers, decompose tensor data, and identify a sparse set of joint biomarkers that are sensitive to temporal cardiac variations.

3.3.2 Modeling Approaches

The ECG sensor network monitors the electrical activities of real-world heart in both space and time and leads to the proliferation of patient monitoring signals. Sensor-based data fusion focuses on the identification of spatiotemporal disease processes (i.e., spatial location/size and temporal deterioration of heart diseases). It may be noted that sensor-based data fusion is more concerned about the inverse problem (i.e., going from sensor signals to make inferences about disease-altered cardiac behaviors), but constructing physical-based models is more explanatory of the underlying mechanisms of cardiac systems (i.e., the forward problem). Thus, as shown in Fig. 3.5, our approach is to integrate the sensor-based data fusion (i.e., ECG/VCG signals from the real-world heart) with the physical-based models (i.e., "virtual" heart) for identifying biomarkers and further optimizing the spatiotemporal diagnostics of heart diseases. In other words, we are more concerned about the decision optimization that goes not only inversely from the space–time data to system physics but also forward from system physics to spatiotemporal observations.

Disease-Altered Spatiotemporal VCG Signals Our prior work showed the fact of spatially shifted 3D ECG trajectory due to the locations of myocardial infarctions. As shown in Fig. 3.6, cardiac vector loops of myocardial infarction (red/dashed)

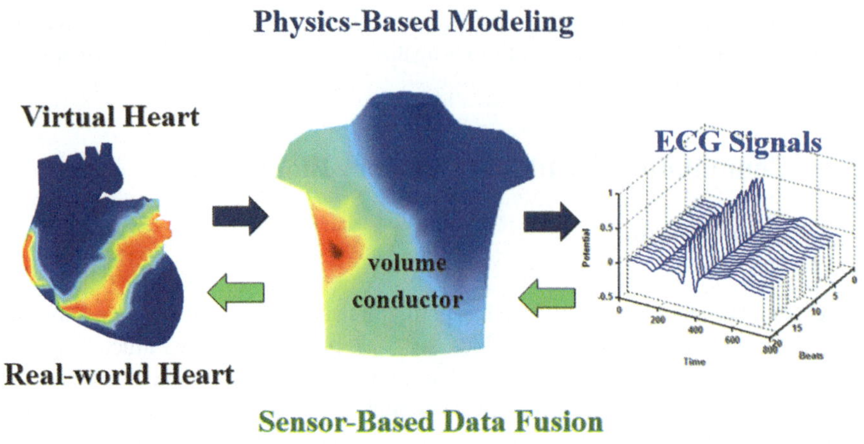

Fig. 3.5 The synergistic fusion of physical-based modeling and sensor-based data fusion for optimizing the spatiotemporal diagnostics of cardiovascular diseases

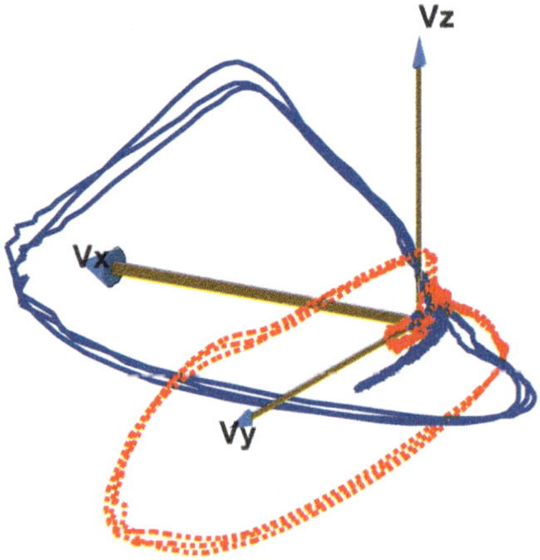

Fig. 3.6 Illustration of 3-lead VCG trajectories of healthy control (blue/solid) and myocardial infarction (red/dashed) [26]

yield a different spatial path from the healthy controls (blue/solid). However, most previous works focused on the patterns (e.g., heart rate, ST segment, QT interval) in the time-domain ECG signals and overlooked the spatiotemporal VCG signals. Our prior studies of sensor-based data fusion revealed the fact of spatially shifted VCG signals due to the locations of myocardial infarctions [8, 25, 26]. We investigated the differences of VCG signals between myocardial infarction (MI) and healthy control (HC) using real-world signals from the PTB database [27] available in the PhysioNet [28]. The statistical hypothesis tests showed significant (p-value<1×10-9) differences in the spatial VCG distributions between MI (368 subjects) and HC

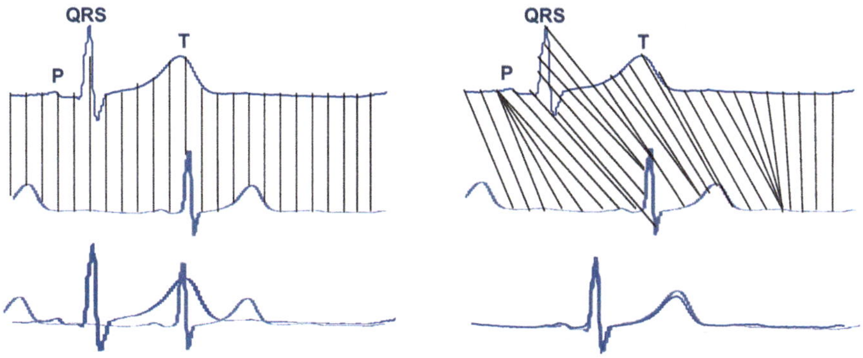

Fig. 3.7 Illustration of 1-lead ECG signals and PQRST alignment

(80 subjects) groups. The predictive model yielded a high sensitivity of 97.28% and a high specificity of 95.00% for the identification of MI subjects.

Network Modeling of Multi-lead ECG Signals It is common to see near-periodic ECG signals, and yet there are variations in shape, amplitude, and phase between ECG cycles (e.g., see Fig. 3.7). Characterization and quantification of such pattern variations provide a great opportunity for monitoring and tracking temporal changes in ECG waveforms. Figure 3.7 shows P, QRS, and T loops of 1-lead ECGs that are not aligned and aligned respectively for two subjects. This alignment is indispensable in terms of correct comparisons of electrical activities from the corresponding heart chambers. In other words, ventricular depolarization (i.e., QRS loops) should be compared between two subjects (or cycles), instead of incorrect comparison between ventricular depolarization (QRS loop) and atrial depolarization (P loop). However, this example only illustrates the proper alignment of 1-lead ECG signals between two subjects (or ECG cycles). There is also an urgent need to measure the dissimilarities between multi-lead ECG waveforms.

Given two multi-lead ECG waveforms $\{\overrightarrow{\mathbf{v}}_1(t_i); i = 1, \cdots, N_1\}$ and $\{\overrightarrow{\mathbf{v}}_2(t_j); j = 1, \cdots, N_2\}$, the alignment should be performed in the multi-dimensional space (also see Fig. 3.8). This is analogous to identify the warping path g to measure the time-normalized spatial distance between $\overrightarrow{\mathbf{v}}_1(t)$ and $\overrightarrow{\mathbf{v}}_2(t)$) as $\sum_{(t_i,t_j)\in g} \|\overrightarrow{\mathbf{v}}_1(t_i) - \overrightarrow{\mathbf{v}}_2(t_j)\|$. The path $g(i, j)$ connects $(1, 1)$ and (N_1, N_2) in a 2-dimensional square lattice and satisfies the constraints of monotonicity and connectedness. The 2-dimensional square lattice is formed by spatial indices in the multi-dimensional space, as shown in Fig. 3.8. To find the optimal path g_*, an exhaustive search of alignment paths is intractable and computationally expensive. Hence, our prior work solves the problem of spatiotemporal alignments with dynamic programming [29]. This algorithm will start at the initial condition: $g(1, 1) = d(1, 1) = \|\overrightarrow{\mathbf{v}}_1(t_1) - \overrightarrow{\mathbf{v}}_2(t_1)\|$ and the warping window $|i - j| < r$. The

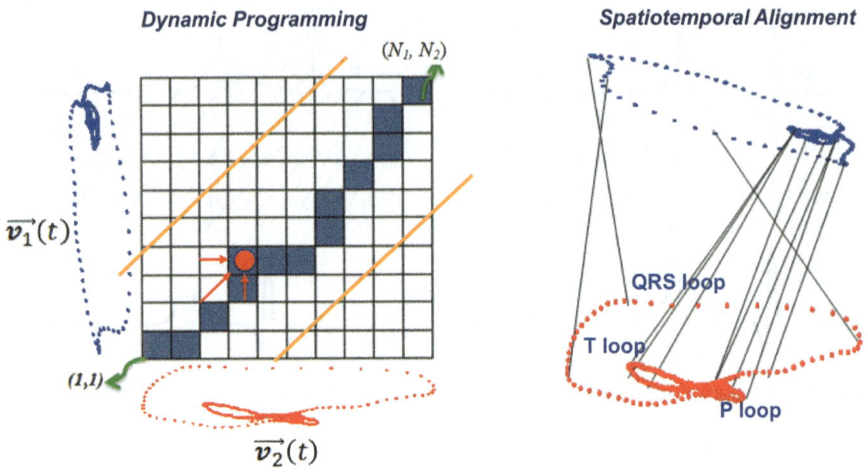

Fig. 3.8 Space–time alignment of 3-lead VCGs

algorithm will search forward as

$$g(i, j) = \min\left(g(i, j-1) + d(i, j), g(i-1, j-1) + +2d(i, j), g(i-1, j)d(i, j)\right).$$

The time-normalized spatial distance is calculated as: $\Delta(\vec{\mathbf{v}}_1, \vec{\mathbf{v}}_2) = \frac{g(N_1, N_2)}{N_1+N_2}$. If there are multi-lead ECG waveforms from M cycles (or subjects), then Δ becomes an $M \times M$ adjacency matrix that represents pairwise dissimilarities.

However, the adjacency matrix Δ cannot be directly used as biomarkers to describe temporal cardiac variations. Notably, the measure of Euclidean distance is not directional and cannot distinguish the distances that are equal in magnitudes but along different spatial directions. Therefore, we propose to transform the adjacency matrix Δ into feature vectors that preserve the warping distances between ECG waveforms. Now the question becomes "how to bridge from ECG waveforms to the adjacency matrix Δ to feature vectors (a node's spatial location in the network)?" The feature vectors will approximately preserve the adjacency matrix Δ between ECG waveforms. This is similar to a network problem, i.e., how to reconstruct the nodes' location in a high-dimensional space if the adjacency matrix Δ is known.

Hence, our prior work leveraged the idea of spatial embedding to represent ECG waveforms as the nodes in a high-dimensional network [26]. Let δ_{ij} denote the dissimilarity between the ith and jth ECG waveforms in $M \times M$ adjacency matrix Δ, and $\mathbf{x}_i$ and $\mathbf{x}_j$ denote the ith and jth nodes in a high-dimensional network. Then, the objective function of spatial embedding will be: $\min \sum_{i<j}(\|\mathbf{x}_i - \mathbf{x}_j\| - \delta_{ij})$; $i, j \in [1, M]$, where $\|\cdot\|$ is the Euclidean norm. These nodes will be used as feature vectors that recover not only the warping matrix but also directional differences between ECGs. Further, we leveraged the dynamic programming algorithms to optimize the embedding of ECG waveforms as network nodes that preserve the adjacency

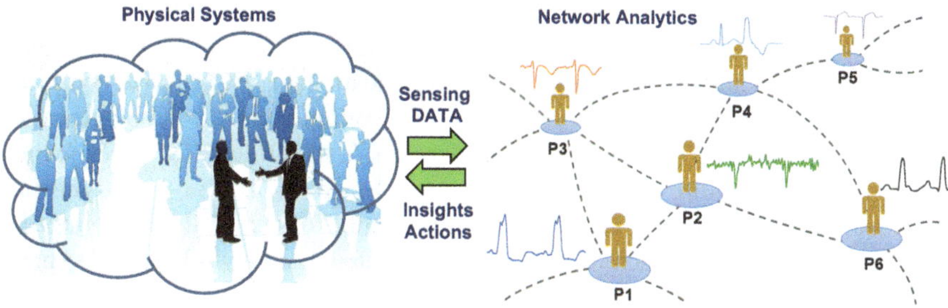

Fig. 3.9 The Internet of Hearts—Cardiac dynamics of human subjects are reflected in the cyber space through data-driven information processing, modeling, and simulation. Analytics in the cyber space exploits the acquired knowledge and useful information from data for the feedback loop of optimal actions (or control schemes) to the physical systems, e.g., inferring dynamic system properties from sensor observations

matrix Δ. An online learning strategy can also be designed to sequentially embed an incoming waveform in the existing network. To this end, dynamic network modeling represents temporal variations in multi-lead ECG signals, and the deviation of node locations characterizes the varying pattern of ECG waveforms.

The Extension to Internet of Hearts If ECG sensing is deployed in the large-scale IoT context, large amounts of data streams will rise. This, in turn, provides an unprecedented opportunity to study disease mechanisms at the population level. Realizing full potential of IoT infrastructure and data calls upon the design of analytical methodologies to extract useful information about abnormal cardiac activity (i.e., from the individual level to the population level). As a result, data-driven knowledge could bring fundamental improvements to medical decision making.

Figure 3.9 shows the new framework of an Internet of Hearts (IoH), which leverages an Internet infrastructure and cloud database to connect a large number of individuals via the ECG sensing. This IoH framework leads to overwhelmingly big data in the individual level and in the population level. Because this IoH network is not available before, most of the existing works are less concerned about the development of new information processing capability in the IoH context. Indeed, there is an urgent need to investigate big data analytics (e.g., personalized-level monitoring and population-level informatics) in the large-scale IoH context. Our prior work developed a multi-level network model to monitor and identify abnormal cardiac activities for IoT-enabled smart management of heart health [30, 31]. An attractive feature is the provision of stochastic learning approaches to optimize the embedding of cardiac signals into a beat-to-beat (B2B) network model, as well as a patient-to-patient (P2P) network model. This multi-level network modeling approach is embodied by the following three components:

- Personalized-level beat-to-beat (B2B) network:
  Wearable sensing provides a higher level of flexibility to monitor long-term ECG signals from an individual. Within a heartbeat, the waveform of ECG signals

varies significantly at different stages of cardiac operations (e.g., excitation from SA node, atrial depolarization and repolarization, as well as ventricular depolarization and repolarization). Among heartbeats, ECG waveforms are similar to each other but with variations. For example, the second heartbeat of patient P4 (see Fig. 3.9)) shows significant variations in the QRS waves, which is a sign of premature ventricular contraction. Stochastic beat-to-beat dynamics can be characterized with the B2B network model for personalized monitoring of heart health, where each heartbeat can be treated as a network node and node attributes are ECG patterns within this heartbeat.

- Population-level patient-to-patient (P2P) network:
  This network model can be further extended to the population level. Note that there are similarities and dissimilarities in ECG patterns among subjects (see Fig. 3.9). As such, this allows for the design and development of patient-to-patient (P2P) network model, where each node represents an individual and node attributes are the salient ECG patterns from this individual. This P2P network provides macro-level informatics for the population, while the B2B network is more concerned about the micro-level details for an individual. Multi-level network modeling helps clinicians identify communities that share similarity in cardiac activity, study disease mechanisms in each community, and pinpoint a subject in one of the communities.
- Parallel graph analytics:
  Multi-level network modeling involves the P2P network and B2B network, thereby forming the new network of networks for ECG analysis at both the individual level and the population level. Nonetheless, large-scale network modeling is computationally expensive due to population size and data volume. Traditional serial-computing schemes are limited in the ability for efficient network representation and real-time analytics in the IoH setting. The power of IoH lies in the inclusion of more subjects for the formation of communities. Our prior work presents the design and development of parallel graph algorithms for efficient modeling and optimization of the large-scale IoH for healthcare applications.

Tensor Decomposition for Joint Biomarker Identification After the online extraction of ECG biomarkers, we often have a temporal log of ECG, context, and blood test variables (i.e., tensor data). Thus, there is a need to investigate the structure and correlation in the high-dimensional tensor data and identify a sparse set of joint biomarkers sensitive to temporal variations in the cardiac process. Traditionally, principal component analysis (PCA) is a common approach for dimensionality reduction but is not applicable here for high-order tensors. Hence, our prior work leverages a new approach of constrained tensor decomposition to extract uncorrelated features from tensor data [12, 13]. This idea is originated from the recent literature on multilinear subspace learning in image and video processing. In contrast, our objective is to identify and extract joint biomarkers from tensor-form data generated by heterogeneous modalities of medical sensing. A significant

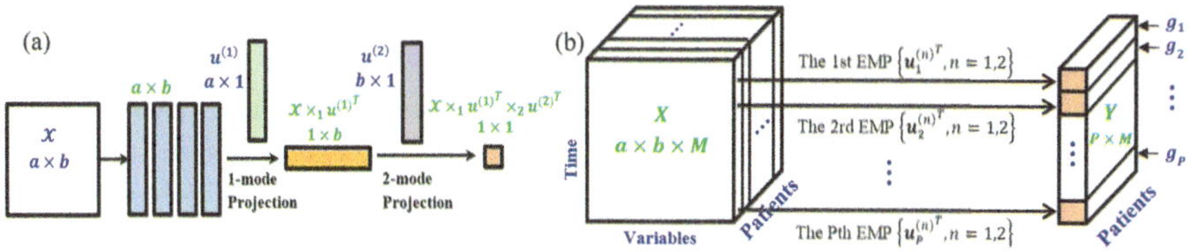

Fig. 3.10 (**a**) An example of EMP of order-2 tensors; (**b**) An Example of P EMPs of order-3 tensors [12, 13]

difference between image/video processing and healthcare applications lies in data heterogeneity. Image/video data are homogeneous and synchronized in time, but cardiac sensing can involve a heterogeneous set of variables that are asynchronized and incomplete.

As shown in Fig. 3.10, multiple elementary multilinear projections (EMP) can project a tensor $\mathcal{X} \in \mathbb{R}^{I_1 \times \cdots \times I_N}$ to a scalar y through the N projection vectors as

$$y = \mathcal{X} \times_1 \mathbf{u}^{(1)^T} \times_2 \mathbf{u}^{(2)^T} \cdots \times_N \mathbf{u}^{(N)^T},$$

where $\mathbf{u}^{(N)}$ is the nth projection vector and $\times_n$ is the nth-mode multiplication. Figure 3.10a illustrates the EMP of an order-2 tensor to a scalar $y = \mathcal{X} \times_1 \mathbf{u}^{(1)^T} \times_2 \mathbf{u}^{(2)^T}$. The use of P EMPs will project a tensor object $\mathcal{X}$ to a vector $\mathbf{y} \in \mathbb{R}^P$ as

$$\mathbf{y} = \mathcal{X} \times_{n=1}^{N} \left\{ \mathbf{u}_p^{(n)^T}, n = 1, 2, \cdots, N \right\}_{p=1}^{P}.$$

Figure 3.10b shows the projection of order-3 tensor $\mathcal{X}$ to a 2-dimensional matrix $\mathbf{Y}$. The $\mathbf{y}_m(p)$ is the projection of mth sample $\mathcal{X}_m$ by EMPs:

$$\mathbf{y}_m(p) = \mathcal{X}_m \times_{n=1}^{N} \left\{ \mathbf{u}_p^{(n)^T}, n = 1, 2, \cdots, N \right\}.$$

As shown in Fig. 3.10b, $\mathbf{g}_p$ denotes the pth coordinate vector and $\mathbf{g}_p(m) = \mathbf{y}_m(p)$. In order to maximize the variance of projections and extract uncorrelated features, a constraint function is imposed on the pth EMP:

$$\left\{ \mathbf{u}_p^{(n)^T}, n = 1, 2, \cdots, N \right\} = \underset{\mathbf{u}_p^{(n)^T}}{\arg\max} \operatorname{Var}(\mathbf{g}_p)$$

$$\text{s.t.} \quad \mathbf{u}_p^{(n)^T} \mathbf{u}_p^{(n)} = 1, p = 1, 2, \cdots, P, \quad \text{and} \quad \mathbf{g}_p^T \mathbf{g}_p = 0, \text{for all } p, q, p \neq q.$$

Further, we utilized a sequential variance maximization algorithm to solve the objective function subject to the normalization and orthogonal constraints and evaluate the proposed tensor decomposition with real patient data extracted from

Multiparameter Intelligent Monitoring in Intensive Care (MIMIC) II Clinical Database. Note that the dataset is in the form of $14 \times 12 \times 8000$ tensor for clinical variables, where 14 is the number of clinical variables (i.e., Lactate, BUN, Creatinine, Glucose, HCO3, Platelets, pH, etc.), 12 is the number of temporal samples (1 sample per 4 h for 48 h), and 8000 is the total number of patients. We used the EMPs to transform and reduce the order-3 tensor data into a low-dimensional set of sparse features (i.e., uncorrelated and orthogonal). Experimental results demonstrate great potentials of tensor decomposition for joint biomarker identification and thereby improve the capability of medical decision making.

3.3.3 Summary

This section presents a dynamic network approach to model and characterize temporal variations of multi-lead ECG waveforms (i.e., either cycle-vs-cycle or subject-vs-subject). That, in turn, facilitates the extraction of ECG biomarkers sensitive to cardiac variations instead of extraneous noises. This section also reviews a constrained tensor decomposition to extract a sparse set of uncorrelated biomarkers from the high-dimensional tensor of heterogeneous clinical data (e.g., ECG, context-sensing, and blood test variables).

3.4 Spatiotemporal Monitoring and Modeling

3.4.1 Introduction

The deployment of body-area network of ECG sensors provides the noninvasive electrocardiographic image (ECGI) [11], also called body-surface potential mapping (BSPM). As shown in Fig. 3.11, ECGI data are spatially and temporally varying

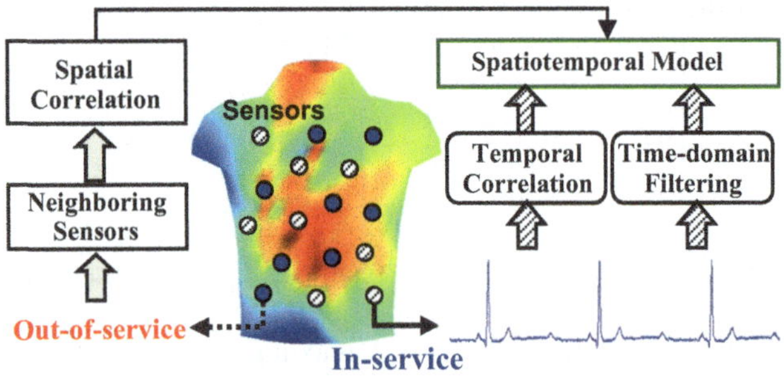

Fig. 3.11 The diagram of spatiotemporal model of body-surface potential mapping

that pose significant challenges on sensor-based modeling and monitoring of cardiac operational dynamics. Statistical control charts are commonly used for quality control and early detection of process deviations in manufacturing processes [32–34]. Statistical monitoring has yielded enormous successes for quality improvements in automotive industry since early 1920s. However, conventional control charts focus more on a low-dimensional set of quality variables (e.g., geometric measurements) and require the assumption of linear model and normal distribution. In contrast, ECGI signals pertinent to human heart are often nonlinear and do not follow Gaussian distribution. Hence, conventional methods are not directly applicable for cardiac monitoring. For example, Hotelling T^2 control charts use the test statistic $T^2 = (\mathbf{Y} - \bar{\mathbf{Y}})^T \boldsymbol{\sigma}^{-1} (\mathbf{Y} - \bar{\mathbf{Y}})$ to monitor the multivariate deviation of a vector of quality variables $\mathbf{Y} = [y_1, y_2, \ldots y_p]$, where $\bar{\mathbf{Y}}$ and $\boldsymbol{\sigma}$ are the sample mean and covariance matrix. Although proven to be effective in many industrial settings, multivariate normal distribution is too rigid to characterize the temporal variation of ECGI signals.

In addition, there are approximately 100s of ECG electrodes used in conventional ECGI systems. Each sensor has to maintain tight contacts with the skin surface for reliable readings. It is not uncommon that some sensors may be temporarily out of service, which brings a high level of data uncertainty. The quality of sensor signals is very critical to realize the system functionality. This hardware-posed constraint makes current ECGI systems extremely uncomfortable and hampers the wide application of ECG imaging from research laboratory to healthcare settings. However, if the hardware requirements are reduced, new models need to be developed to assure the quality of ECGI data. Given the stochastic electrode-skin contacts, ECG sensor network will capture cardiac electrical activities intermittently at the dynamically varying locations of electrodes (see Fig. 3.11). This new problem has received few attention in the past, partly due to the traditional designs that sacrifice the wearability for the data quality, and partly to the fact that traditional algorithms only consider the spatial or temporal estimation but not both. Therefore, our previous works propose a new concept of **"Stochastic Sensor Network,"** which *"allows a subset of sensors at varying locations within the network to transmit dynamic information intermittently"* [35–37]. This section presents a practical spatiotemporal monitoring solution by real-time computing, continual updating, and harnessing of stochastic sensing data.

3.4.2 Modeling Approaches

New modeling approaches are urgently needed to support the unconventional design of stochastic sensor networks. Thus, we first develop a graphical user interface (GUI) to visualize the ECGI data and evaluate a preliminary version of estimation algorithms using the gathered real-world ECGI from ECG image databases [38], available in PhysioNet [28]. The ECG image database consists of 4 sets of ECG imaging data at 352 torso-surface nodes. The ECG signal at each site contains a

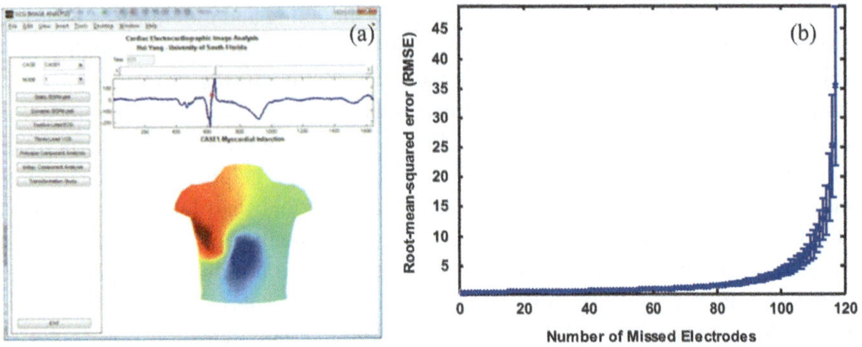

Fig. 3.12 (**a**) Screenshot of the developed graphical user interface software for ECG Imaging visualization, also see YouTube demo in this link: https://youtu.be/KaCz5gx6pIA. (**b**) The variation of RMSE error vs. the number of non-contact electrodes. The simulation uses real-world datasets to mimic the missing recordings at the locations of randomly chosen n electrodes out of 120

single averaged PQRST complex sampled at 2 kHz. The ECG image database also provides detailed anatomic locations of all 352 ECG nodes on the torso surface. As shown in Fig. 3.12a, we developed graphical user interface (GUI) software to visualize the ECG images on a finite element torso.

Because the stochastic nature of ECG network design employs a variable degree of contacts between skin and sensors, we utilized the available ECG imaging datasets to simulate scenarios of intermittent or completely missing recordings at some sensor sites. The simulation mimics the conditions of completely missing recordings at the locations of randomly chosen n electrodes out of 352. The question is: "How can we develop a function to estimate the signals at these random n locations?" As shown in Fig. 3.12b, we used a linear spatial correlation matrix to estimate the missing space–time variables. The results show that the error of estimation (measured in terms of root-mean-square error (RMSE)) rises when the number of non-contact electrodes increases. It may be noted that the RMSE error has a significant increase when the number of non-contact electrodes is greater than 80. But the error is in the plateau phase, i.e., smaller if there are <60 ECG electrodes lose contact with skin. The information redundancy in ECG imaging strongly suggests the viability of stochastic ECG sensor network for the acquisition of ECGI signals. In other words, it is not necessary to attach all electrodes to the surface of skin in the presence of a relatively large set of electrodes. Collectively, these simulation experiments confirm that new spatiotemporal modeling approaches can be developed to realize the unconventional design of stochastic sensor network and harness information from stochastic sensor-skin contacts.

Spatial Modeling of ECG Image As shown in Fig. 3.13, ECG images vary over time in the body surface when the heart undergoes different stages of operations. At a given time, ECG image shows nonlinear and nonstationary patterns in the spatial domain (i.e., the body surface). If two locations are spatially closer to each other

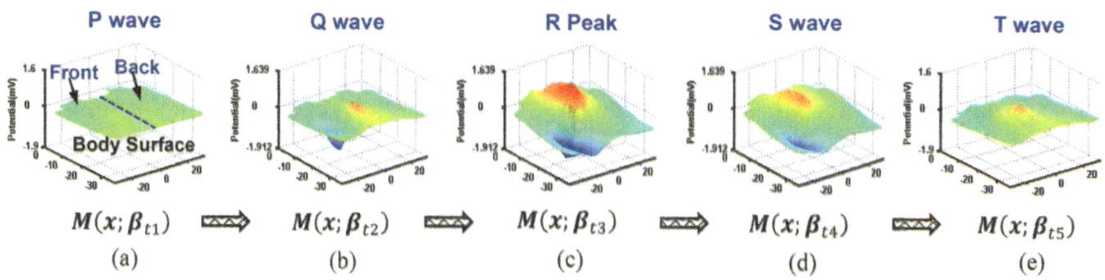

Fig. 3.13 Time-varying ECG imaging surfaces at the locations of: (**a**) P wave; (**b**) Q wave; (**c**) R peak; (**d**) S wave; and (**e**) T wave

on the ECG image, they tend to have a stronger correlation. The key to spatial modeling of ECG image is to characterize and estimate the spatial correlation between neighboring sensors. In the case of stochastic sensor-skin contacts, spatial models are critical to predict the ECGI data values that are varying in the body surface. Figure 3.13 shows that ECG images are spatially varying in the front and back of the body at a specific time point and are also temporally varying from P, QRS, T waves with respect to time. In order to capture the spatial correlation in ECG images, we design the spatial model in the form as follows:

$$\mathbf{Y}(s) = \mathbf{M}(s; \boldsymbol{\beta}) + \epsilon(s) = \sum_{i=1}^{N} w_i(s) \boldsymbol{f}_i'(s) \boldsymbol{\beta}_i + \epsilon(s). \tag{3.1}$$

This model does not involve the time index t yet because it is focused on the ECG image at a specific time point. Here, $\mathbf{Y}(s)$ is the ECGI data value at the spatial location s on the body surface, $\boldsymbol{\beta}_i = (\beta_{i1}, \cdots, \beta_{ip})'$ is a vector of model parameters, N is the total number of components, and $\epsilon(s)$ is Gaussian random noises. The $w_i(s)$ is a non-negative weighting kernel function, i.e.,

$$w_i(s) \propto |\textstyle\sum_i|^{-\frac{1}{2}} \exp\left\{-\frac{1}{2}(s - \boldsymbol{\mu}_i)' \textstyle\sum_i^{-1} (s - \boldsymbol{\mu}_i)\right\}.$$

The $\boldsymbol{f}_i(s) = (f_{i1}(s), \cdots, f_{ip}(s))'$ is a set of known basis functions. In other words, this spatial model can be viewed as a weighted mixture of basis function regressions. For one realization of spatiotemporal process, we observe $\mathbf{Y} = (Y(s_1), \cdots, Y(s_n))$ at spatial locations $s_1, \cdots, s_n$ on the body surface. If we define $\boldsymbol{w}_i = (w_i(s_1), \cdots, w_i(s_n))$ as the vector of weights for kernel i and $\boldsymbol{X}_i = (\boldsymbol{f}_{i1}(s), \cdots, \boldsymbol{f}_{ip}(s))$ as the design matrix for the ith mixture component, this spatial model can be rewritten as basis function regression: $\mathbf{Y} = \boldsymbol{X\beta} + \epsilon$, where $\boldsymbol{X} = (\text{diag}(\boldsymbol{w}_1)\boldsymbol{X}_1, \cdots, \text{diag}(\boldsymbol{w}_N)\boldsymbol{X}_N)$ and $\boldsymbol{\beta} = (\boldsymbol{\beta}_1', \cdots, \boldsymbol{\beta}_N')'$. As a result, model parameters can be estimated from available ECGI data. Once the spatial model is constructed and estimated with large amounts of data, spatially missing value $\mathbf{Y}(\hat{s})$ (i.e., due to stochastic sensor-skin contacts) can be predicted as $\boldsymbol{X}\hat{\boldsymbol{\beta}}$ for the location $\hat{s}$.

In general, the ECG image at a given time point is nonlinear and nonstationary. This kernel-weighted regression model offers the capability and flexibility to model the spatial variations of ECG images. There are also research questions to be investigated in the future work. For example, instead of Gaussian kernels, can different kernel forms be selected or optimized for $w_i(s)$ in Eq. (3.1) to make the model more compact? In addition, can the form of kernel function $w_1(s)$ be the same or different from that of $w_2(s), \cdots, w_i(s)$ to achieve a higher goodness of fit to the data. Thus, it is necessary to design and develop kernel functions for the specific application in the context of ECGI sensing so as to increase the model sparseness and convergence. If the kernel functions are optimized, then the spatial model could converge faster and be sparser. In other words, a sparse and finite number of terms N is sufficient to yield superior performance in the approximation. The computation overhead will also be significantly reduced.

Spatiotemporal Model of ECG Image ECGI is not only spatially varying but also temporally propagating in the body surface when the heart is functioning over time. To account for spatiotemporal dynamics, we further redesign the spatial model in Eq. (3.1) to include temporal components. The basic idea is to allow model parameters to vary over time, i.e., $\boldsymbol{\beta}_t = (\beta_{t1}, \cdots, \beta_{tp})'$ over the time index t. As shown in Fig. 3.13, as cardiac electrical activities are varying in both space and time, time-varying parameters enable spatial ECGI data $\mathbf{M}(x; \boldsymbol{\beta}_1), \mathbf{M}(x; \boldsymbol{\beta}_2), \cdots, \mathbf{M}(x; \boldsymbol{\beta}_t), \cdots$ to vary with respect to time.

Our prior work designed and developed a sparse particle filter [36] as a nonlinear state space model to link the evolving parameters $\boldsymbol{\beta}_t$ over time. The sparse particle filter recursively estimates and predicts the time-varying parameters via the evolution equation:

$$\boldsymbol{\beta}_t = \mathbf{G}_t \boldsymbol{\beta}_{t-1} + \boldsymbol{\omega}_t, \quad \boldsymbol{\omega}_t \sim N(0, \mathbf{Z}_t), \tag{3.2}$$

where parameters $\boldsymbol{\beta}_t$ and $\boldsymbol{\beta}_{t-1}$ are linked by the evolution matrix $\mathbf{G}_t$ and $\boldsymbol{\omega}_t$ is process noise with zero mean and covariance matrix $\mathbf{Z}_t$. Once the parameters $\boldsymbol{\beta}_t$ become available, spatial variations of ECGI in the body surface at a given time can then be defined as

$$\mathbf{Y}_t = \mathbf{X}_t \boldsymbol{\beta}_t + \boldsymbol{\epsilon}_t, \quad \boldsymbol{\epsilon}_t \sim N(0, \mathbf{V}_t), \tag{3.3}$$

where $\mathbf{Y}_t$ is the observed ECG image at time t and $\boldsymbol{\epsilon}_t$ is the measurement noise with zero mean and covariance matrix $\mathbf{V}_t$. The sparse particle filter serves as a new spatiotemporal monitoring framework for the ECGI data on the body surface. At a given time point, this monitoring framework estimates the state of model parameters $\boldsymbol{\beta}_t$ once the data become available. Then, the temporal predictor Eq. (3.2) is responsible for predicting the model parameters $\boldsymbol{\beta}_{t+1}$ at the next time point. This, in turn, projects forward (over time) the current state to the next state and estimates error covariance to obtain the priori estimates for $\boldsymbol{\beta}_{t+1}$ (and the spatial ECGI $\mathbf{Y}_t$) at the next time step.

When the prediction $\boldsymbol{\beta}_{t+1}$ becomes available, the spatial update Eq. (3.3) provides the feedback. In other words, new measurement data of ECG image are integrated with the priori estimate to obtain an improved posteriori estimate. This process iteratively alters between temporal prediction and spatial update. As a result, the spatiotemporal monitoring framework leverages previous posteriori estimates to project or predict the new priori estimates and then update body-surface ECGI over time. This recursive filtering framework makes it practical to implement the state space model to link time-varying model parameters, as well as model the spatial variations of ECGI data. An attractive feature is to combine the spatial model with temporal-filter model in this spatiotemporal monitoring of ECGI data. See more technical details in our previous publications [35–37].

3.4.3 Summary

This section presents a new concept of "stochastic sensor network" for ECG imaging. Spatiotemporal models are designed and developed to allow for the optimization of stochastic sensor network while maintaining the data integrity. Our prior work is the first of its kind to utilize advanced software algorithms (i.e., spatiotemporal modeling of ECGI data) to relax the strict criterion of hardware designs (i.e., tight electrode-skin contact) in ECGI systems. The network of ECG electrodes brings the proliferation of spatiotemporal cardiac monitoring signals. New spatiotemporal models provide viable approaches for real-time integration and imputation of high-dimensional BSN data that spatially temporally distributed, asynchronous, and incomplete.

In this section, we have also discussed research questions to be investigated in the future work. For e.g., (1) should the same Gaussian kernel or different kernels be selected or optimized for $w_i(s)$ in Eq. (3.1)? (2) how to correlate space–time patterns hidden in ECGI data with the spatial localization and temporal deterioration of myocardial infarctions? (3) how to derive personalized model characteristics that can serve as sensor-based biomarkers of the change in cardiac conditions manifested in space–time ECGI data. In general, modeling ideas and concepts presented in this section can be generally applicable to spatiotemporal data modeling and imputation in stochastic sensor networks.

3.5 Automatic Disease Detection from ECG Signals

3.5.1 Introduction

Routine use of ECG signals is a common method to examine electrical activities of the heart, which is the only effective way to detect silent heart attacks. In current practice, cardiac arrhythmia is generally identified by checking ECGs with naked eyes, which may be subject to human errors and require extensive human resources. Effective computational methods are urgently needed for inexpensive data-driven detection of arrhythmia. Various statistical algorithms have been proposed to investigate ECG signals such as dynamic time warping [39], wavelet analysis [40], and Bayesian network [41]. Additionally, various deep neural network (DNN) methods including convolutional and recurrent neural networks (i.e., CNNs, RNNs) have been developed for cardiac diagnosis and outperformed traditional statistical methods [42]. Note that DNN models based on supervised learning require a large number of labeled data for effective training. However, annotating ECG cycles requires medical experts to manually examine the signals and provide a label to each pattern, which is a labor-intensive process. Moreover, it is impractical to collect enough balanced data for different types of disease-altered signals to satisfy the requirement for sufficient supervised training because different arrhythmias can have highly diverse occurrence rates. As a result, semi-supervised learning has been increasingly investigated for ECG analysis [43–45]. However, most of the existing semi-supervised methods are limited to binary classification, i.e., differentiating abnormal ECGs from normal ones, and are not able to identify different types of arrhythmias. Thus, advanced deep learning models are in high demand to efficiently cope with the data-lacking and imbalanced data issues for both accurate anomaly detection and disease identification.

3.5.2 Two-level DNN with Generative Adversarial Network

Our prior work has developed a two-level DNN model with generative adversarial network (GAN) to cope with the data-lacking and imbalanced data issues for effective auto-detection of different types of arrhythmias [46]. This framework consists of two levels of modeling: memory-augmented deep auto-encoder with GAN (MadeGAN) for the first-level anomaly detection, and transfer-learning- and multi-branching-enhanced second-level multi-class classification to identify different types of arrhythmias.

First-Level Model: MadeGAN for Anomaly Detection

Cardiac events typically happen intermittently. Continuous cardiac sensing often generates a much larger dataset for normal ECG signals than abnormal ones. The first-level MadeGAN is developed to address such imbalanced issues, which will be trained by normal signals and output anomaly scores to differentiate abnormal ECGs from normal ones. Note that AutoEncoder (AE) is a commonly used deep learning method for anomaly detection. The AE consists of an encoder and a decoder and leverages a feedforward DNN for representation learning. However, because of the powerful approximation capability of DNN, the traditional AE trained by normal samples with diverse patterns tends to generalize well and may reconstruct anomalous signals with very small reconstruction errors, resulting in sub-optimal performance for anomaly detection. To address this problem, our MadeGAN model incorporates a memory module and the adversarial training technique into the AE for robust anomaly detection, which includes the following four key components:

(1) **Encoder**: The encoder maps the original input data $\boldsymbol{x}$ to a latent space with a reduced dimensionality z using function $f_e(\cdot; \boldsymbol{\theta}_e)$, i.e., $z = f_e(\boldsymbol{x}; \boldsymbol{\theta}_e)$ to capture the key information in the input signal. The latent vector z will then serve as a query to retrieve relevant elements in the memory module.
(2) **Memory module**: The memory module is designed to store representative latent ECG features, which is formed by a matrix $\boldsymbol{\Omega}$. Specifically, the memory module extracts the prototypical vectors in $\boldsymbol{\Omega}$ similar to the query vector z to formulate the representation vector $\hat{z}$, which will then be fed to the decoder to map back to the input domain from the latent domain.
(3) **Decoder**: The decoder network is designed to map the representative latent vector $\hat{z}$ back to the original input space and output the reconstruction of $\boldsymbol{x}$ using the function $f_d(\cdot; \boldsymbol{\theta}_d)$: $\hat{\boldsymbol{x}} = f_d(\hat{z}; \boldsymbol{\theta}_d)$. The similarity between $\hat{\boldsymbol{x}}$ and $\boldsymbol{x}$ will serve as an indicator of the anomaly degree of input signal $\boldsymbol{x}$.
(4) **Adversarial training**: To further enlarge the gap in reconstruction errors between abnormal and normal signals, the framework of GAN is leveraged to encode adversarial training into AE. The GAN consists of two networks: a generator G created to generate synthetic signals that are similar to the real data and a discriminator D designed to classify the signal as real or fake (artificial). The competitive objectives of D and G enable the GAN to generate high-quality synthetic samples. We propose to adopt the AE with memory module (MemAE) as the generator of GAN to design the MadeGAN for first-level anomaly detection. As such, the adversarial training will push the generator MemAE to improve in synthesizing realistic reconstructions from normal input signals, further enlarging the gap in the reconstructions from normal and anomalous signals.

Second-Level Model: Transfer-Learning- and Multi-Branching-Enhanced Classification

The quality and quantity of training data significantly impact the performance of DNN-based classification. However, data associated with arrhythmia are much less than normal data due to the expensive process for data collection and annotation. Moreover, the diverse occurrence rates of different diseases often lead to imbalanced sample sizes for different arrhythmia classes. To solve the aforementioned issues, our second-level model first leverages transfer learning to take advantage of the large normal ECG signals to improve the classification performance [47]. Specifically, we adopt the trained discriminator D in the first-level MadeGAN to improve the second-level multi-class classification. The trained discriminator D serves as a feature extractor to extract key features from abnormal ECGs, which will then be fed

Table 3.1 The comparison of classification performance between our two-level hierarchical model and existing classification methods [46]

Authors	Recall	Precision	Accuracy
Proposed method	0.964	0.967	0.967
Li et al. [52]	0.647	0.475	0.946
Acharya et al. [53]	0.960	0.915	0.940
Mousavi et al. [54]	0.831	0.923	0.987
Kachuee et al. [55]	–	–	0.934

into a shallow 1D CNN classifier for the identification of different arrhythmias. Note that the network parameter in the adopted discriminator D will be frozen (kept as a constant set) during the second-level training. Moreover, we incorporate the multi-branching (MB) module [48–50] to the 1-D CNN to minimize the potential bias induced by the class imbalance problem for accurate classification. By integrating the MB architecture and the transfer learning technique with the 1D CNN, the second-level model overcomes the data-lacking and imbalanced data issues and is able to enhance the performance of data-driven disease identification.

We implemented the two-level hierarchical framework to analyze the ECG signals from the MIT-BIH database [51]. Experimental results demonstrate the 2-level DNN significantly outperforms existing commonly used methods in terms of different performance scores as shown in Table 3.1.

3.5.3 Summary

This section presents a GAN-based two-level DNN framework for ECG signal analysis and arrhythmia identification. The first-level MadeGAN is constructed by integrating the AutoEncoder with a memory module and adversarial training for detecting abnormal cardiac conditions. The second-level classification framework is designed by incorporating transfer learning and multi-branching architecture into a 1D CNN to classify different types of arrhythmia. This two-level framework effectively solves both the data-lacking and imbalanced issues in data-driven modeling for disease detection.

3.6 Characterization of Myocardial Infarction Using Inverse ECG Modeling

3.6.1 Introduction

The ECG is an integrated measure of the dynamic electrical activity of the heart projected on the body surface, which is a macroscopic sum of all cardiac electrical sources. The detailed features of the spatiotemporal cardiac electrodynamics become blurred when projecting from the heart to the torso surface. As such, body-surface ECG has diagnostic limitations especially in the accurate localization and characterization of abnormal regions within the heart [56]. Precision cardiology calls upon advanced models to quantitatively estimate disease-altered cardiac electrical activities from body-surface sensor recordings (e.g., ECGs, BSPMs), which is known as the inverse ECG problem [57]. Being able to accurately estimate the electrical activity within the heart from ECGs or BSPMs is extremely useful and will facilitate the paradigm shift from invasive cardiac diagnostics to noninvasive, painless, and inexpensive precision cardiology.

Inverse ECG modeling aims to reconstruct cardiac electrodynamics from body-surface sensor recordings (i.e., BSPM) in a noninvasive way. Specifically, BSPM $\boldsymbol{y}(s,t)$ is related with heart-surface potential (HSP) $\boldsymbol{u}(s,t)$ according to $\boldsymbol{y}(s,t) = R_{BH}\boldsymbol{u}(s,t) + \epsilon$, where s and t denote the spatial and temporal coordinates, respectively, and R_{BH} is the transfer matrix, which can be derived according to the divergence theorem and Green's second identity using boundary element method. Unfortunately, the inverse ECG problem involves ill-posed equation systems to solve numerically, where a small measurement noise in the input may introduce a high instability in the prediction. A variety of methods have been developed to cope with the ill-conditioned issue and achieve robust inverse solutions by adding proper spatial or/and temporal regularization such as Tikhonov methods [58], L1 regularization [59], spatiotemporal regularization [60], and electrophysiological model-enhanced regularization [61]. This section introduces our prior efforts on spatiotemporal inverse ECG modeling and the characterization of myocardial infarction (MI) on the heart surface based on the estimated HSP.

3.6.2 *Robust Inverse ECG Modeling*

Distributed ECG systems, e.g., BSPM, employ hundreds of electrodes at various locations on the torso surface to record the dynamic changes in cardiac electrical activities, generating high-resolution mappings of body-surface electric potentials that are projected from the spatiotemporal dynamics within the heart. Reconstructing the electric potentials on the heart surface from BSPMs calls for robust inverse ECG modeling to effectively account for the spatiotemporal data structure and cope with the ill-conditioned issue. To achieve this goal, our prior work has developed a spatiotemporal regularization (STRE) method [62, 63] with the following objective function:

$$\min_{\boldsymbol{u}(s,t)} \sum_{t} \Bigg\{ \|\boldsymbol{y}(s,t) - R_{BH}\boldsymbol{u}(s,t)\|_2^2 + \lambda_s^2 \|\Gamma_s \boldsymbol{u}(s,t)\|_2^2 + \lambda_t^2 \sum_{\tau=t-\omega/2}^{\tau=t+\omega/2} \|\boldsymbol{u}(s,t) - \boldsymbol{u}(s,\tau)\|_2^2 \Bigg\}, \quad (3.4)$$

where Γ_s is a spatial Laplacian operator on a 3D geometry to capture the spatial correlation of the distributed signals, and λ_s and λ_t are the parameters to control the regularization degree in the spatial and temporal domains, respectively. By imposing both spatial and temporal regularization into inverse ECG modeling, optimizing Eq. (3.4) will output a reliable and robust estimation of electrical activities within the heart from body-surface sensor measurements.

Note that the STRE method in Eq. (3.4) focuses on exploiting the spatiotemporal data structure, which does not take full advantage of the physics-based principles governing the cardiac electrical wave propagation. To cope with this limitation, our recent work has developed a physics-constrained deep learning (P-DL) approach to solve the inverse ECG problem [64, 65]. Specifically, the P-DL method incorporates the Aliev–Panfilov model [66], a widely used model to describe the knowledge of cardiac electrophysiology, into deep learning to regularize the network training and achieve a robust inverse ECG solution. The P-DL model is trained by minimizing the following loss function:

$$\mathcal{L}(\boldsymbol{\theta}_{PDL}) = \frac{1}{N_{st}} \sum_t \sum_s \|\boldsymbol{y}(s,t) - R_{BH}\hat{\boldsymbol{u}}(s,t)\|^2 \\ + \frac{1}{N_c} \sum_i ((r_u(\boldsymbol{s}_i, t_i))^2 + (r_v(\boldsymbol{s}_i, t_i))^2 + (r_{bc}(\boldsymbol{s}_i, t_i))^2), \quad (3.5)$$

where N_{st} denotes the number of sensor measurements in BSPM and N_c is the number of collocation points to encode the physics law into the neural network. The first term in Eq. (3.5) stands for the data-driven loss, which measures the discrepancy between the observed BSPM and estimated body-surface signal projected from the estimated HSP $\hat{\boldsymbol{u}}(s,t)$. We call the second term in Eq. (3.5) the physics-based loss, which is leveraged to guarantee the estimated signals follow the physics laws satisfactorily, where $r_u(\boldsymbol{s}_i, t_i)$, $r_v(\boldsymbol{s}_i, t_i)$, and $r_{bc}(\boldsymbol{s}_i, t_i)$ represent the residuals of the partial differential equations in the Aliev–Panfilov model:

$$r_u(\boldsymbol{s}, t) := \frac{\partial \hat{u}}{\partial t} - \nabla \cdot (D\nabla \hat{u}) - k_r \hat{u}(\hat{u} - a)(1 - \hat{u}) + \hat{u}\hat{v} \\ r_v(\boldsymbol{s}, t) := \frac{\partial \hat{v}}{\partial t} - \xi(\hat{u}, \hat{v})(-\hat{v} - k_r \hat{u}(\hat{u} - a - 1)) \\ r_{bc}(\boldsymbol{s}, t) := \boldsymbol{n} \cdot \nabla \hat{u}. \quad (3.6)$$

Table 3.2 compares the performance between STRE and P-DL in solving the inverse ECG problem given different noise conditions (i.e., different σ_ϵ's). The performance metric is the relative error (RE), which is defined as

$$RE = \frac{\|\boldsymbol{u}_m - \hat{\boldsymbol{u}}\|}{\|\boldsymbol{u}_m\|}, \quad (3.7)$$

Table 3.2 Comparison of RE between P-DL and STRE in inverse ECG modeling

Method	$\sigma_\epsilon = 0.01$	$\sigma_\epsilon = 0.05$	$\sigma_\epsilon = 0.1$
STRE	0.219	0.239	0.248
P-DL	0.149	0.152	0.154

where $\hat{\boldsymbol{u}}$ and $\boldsymbol{u}_m$ denote the estimated and true HSP. Taking advantage of the physics-based regularization and powerful learning capability of deep neural network, the P-DL method significantly outperforms the STRE under all the noise conditions.

3.6.3 Characterization of MI on the Heart Surface

The heart-surface electrogram estimated from inverse ECG modeling can be further leveraged to characterize the extent and location of diseased tissue, e.g., MI, in the heart. Note that heart-surface electrograms are represented as time series signals. Time–frequency analysis such as wavelet transformation provides the interpretation of time series signals in both the time and frequency domains simultaneously, enabling multi-dimensional description of local, transient, or intermittent patterns inherent to the data. We utilize the Daubechies wavelet to transform the heart-surface electrogram to capture the MI-altered features. Specifically, the heart-surface signal $\boldsymbol{u}(s, t)$ is decomposed into the running averages A_j and running differences D_j to approximate the original time series and describe the local details of the signal, where $j = 1, 2, \ldots$ denotes the level of decomposition. The wavelet decomposition enables the multi-resolution investigation of $\boldsymbol{u}(s, t)$ at different levels of A_j's and D_j's, which are further clustered by hierarchical clustering for MI localization [67, 68].

We use the real patient data from PhysioNet/Computing in Cardiology Challenge 2007 [69] to evaluate the performance of MI localization. This dataset includes the torso-heart geometry, BSPMs from four MI patients, and GE-MRI images that outline the true MI segments for each patient. The four cases are split into two training and two test cases. Three metrics are used to evaluate the characterization performance: (1) **EPD**—percentage discrepancy between the estimated extent of MI and the true extent obtained from GE-MRI images; (2) **SO**—overlap percentage between estimated MI segments and true segments; (3) **CED**—distance from the estimated MI centroid and the true centroid identified from GE-MRI images. Figure 3.14 shows the comparison of EPD and SO between the ST-iECG method and the traditional iECG approach. Note that our ST-iECG method generates a higher EPD and SO. Additionally, the ST-iECG method provides 0 CED for both case 3 and case 4, while the CEDs from iECG are 1 and 2, respectively. As such, ST-iECG provides a more accurate characterization of MI compared with the traditional iECG approach.

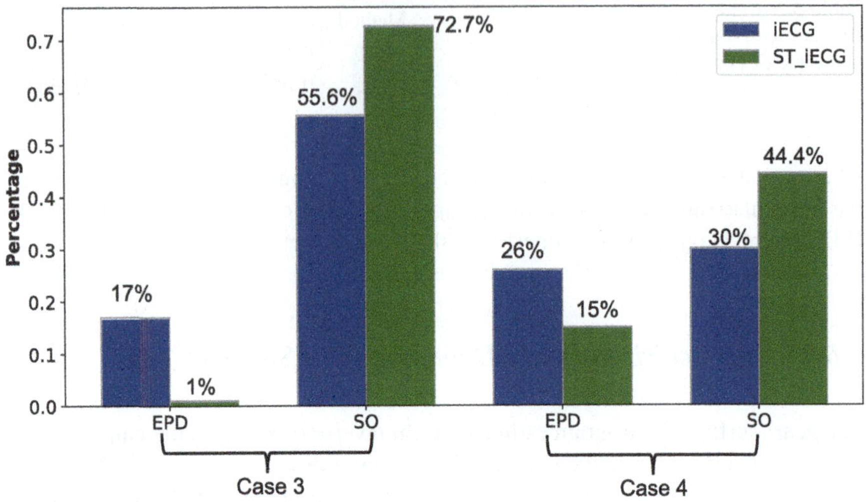

Fig. 3.14 Comparison of EPD and SO between ST-iECG and the iECG model from [69]

3.6.4 Summary

Inverse ECG modeling provides a great opportunity to achieve noninvasive, painless, and inexpensive precision disease diagnosis and characterization. This section presents spatiotemporal regularization and physics-constrained deep learning for robust inverse ECG modeling to reliably estimate heart-surface electrogram from body-surface sensor recordings. The estimated heart-surface signals can be further used to localize the abnormal tissues such as myocardial infarction within the heart.

References

1. H. Yang, S. Bukkapatnam, R. Komanduri, Nonlinear adaptive wavelet analysis of electrocardiogram signals. Phys. Rev. E **76**(2), 026214 (2007)
2. G.D. Clifford, F. Azuaje, P. McSharry, et al., *Advanced Methods and Tools for ECG Data Analysis*. Artech House Boston, vol. 10 (2006)
3. S. Bukkapatnam, R. Komanduri, H. Yang, et al., Classification of atrial fibrillation episodes from sparse electrocardiogram data. J. Electrocardiol. **41**(4), 292–299 (2008)
4. N. Hughes, L. Tarassenko, S.J. Roberts, Markov models for automated ECG interval analysis. Adv. Neural Inf. Proces. Syst. **16**, 1–8 (2003)
5. N.V. Thakor, Y.-S. Zhu, Applications of adaptive filtering to ECG analysis: Noise cancellation and arrhythmia detection. IEEE Trans. Biomed. Eng. **38**(8), 785–794 (1991)
6. Y. Chen, H. Yang, Wavelet packet analysis of disease-altered recurrence dynamics in the long-term spatiotemporal vectorcardiogram (VCG) signals, in *2013 35th Annual International Conference of the IEEE Engineering in Medicine and Biology Society (EMBC)* (IEEE, New York, 2013), pp. 2595–2598

7. G. Liu, H. Yang, Multiscale adaptive basis function modeling of spatiotemporal vectorcardiogram signals. IEEE J. Biomed. Health Inform. **17**(2), 484–492 (2013)
8. H. Yang, Multiscale recurrence quantification analysis of spatial cardiac vectorcardiogram signals. IEEE Trans. Biomed. Eng. **58**(2), 339–347 (2010)
9. Y. Chen, H. Yang, Multiscale recurrence analysis of long-term nonlinear and nonstationary time series. Chaos, Solitons & Fractals **45**(7), 978–987 (2012)
10. H. Yang, S.T. Bukkapatnam, R. Komanduri, Spatiotemporal representation of cardiac vectorcardiogram (VCG) signals. Biomed. Eng. Online **11**(1), 1–15 (2012)
11. Y. Rudy, J.E. Burnes, Noninvasive electrocardiographic imaging. Ann. Noninvasive Electrocardiol. **4**(3), 340–359 (1999)
12. Y. Chen, H. Yang, Heterogeneous postsurgical data analytics for predictive modeling of mortality risks in intensive care units, in *2014 36th Annual International Conference of the IEEE Engineering in Medicine and Biology Society* (IEEE, New York, 2014), pp. 4310–4314
13. Y. Chen, F. Leonelli, H. Yang, Heterogeneous sensing and predictive modeling of postoperative outcomes, in *Healthcare Analytics: From Data to Knowledge to Healthcare Improvement* (2016), pp. 463–501
14. Y. Chen, H. Yang, Self-organized neural network for the quality control of 12-lead ECG signals. Physiol. Meas. **33**(9), 1399 (2012)
15. C. Kan, Y. Chen, H. Yang, Multiscale quality control of telemedicine ECG signal acquisition, in *Telehealthcare Computing and Engineering: Principles and Design* (2011)
16. Y. He, Missing data analysis using multiple imputation: getting to the heart of the matter. Circ. Cardiovasc. Qual. Outcomes **3**(1), 98–105 (2010)
17. H. Liao, J. Sun, Nonparametric and semi-parametric sensor recovery in multichannel condition monitoring systems. IEEE Trans. Autom. Sci. Eng. **8**(4), 744–753 (2011)
18. K. Pelckmans, J. De Brabanter, J.A. Suykens, B. De Moor, Handling missing values in support vector machine classifiers. Neural Netw. **18**(5–6), 684–692 (2005)
19. V.R. Joseph, Limit Kriging. Technometrics **48**(4), 458–466 (2006)
20. V.R. Joseph, Bayesian computation using design of experiments-based interpolation technique. Technometrics **54**(3), 209–225 (2012)
21. C. Cheng, A. Sa-Ngasoongsong, O. Beyca, et al., Time series forecasting for non-linear and non-stationary processes: a review and comparative study. IIE Trans. **47**(10), 1053–1071 (2015)
22. D. Du, H. Yang, A.R. Ednie, E.S. Bennett, Statistical metamodeling and sequential design of computer experiments to model glyco-altered gating of sodium channels in cardiac myocytes. IEEE J. Biomed. Health Inform. **20**(5), 1439–1452 (2015)
23. F. Imani, C. Cheng, R. Chen, H. Yang, Nested Gaussian process modeling for high-dimensional data imputation in healthcare systems, in *Proceedings of the IIE Annual Conference* (Institute of Industrial and Systems Engineers (IISE), New York, 2018), pp. 1312–1317
24. F. Imani, C. Cheng, R. Chen, H. Yang, Nested Gaussian process modeling and imputation of high-dimensional incomplete data under uncertainty. IISE Transactions on Healthcare Systems Engineering **9**(4), 315–326 (2019)
25. D. Dawson, H. Yang, M. Malshe, S.T. Bukkapatnam, B. Benjamin, R. Komanduri, Linear affine transformations between 3-lead (Frank XYZ leads) vectorcardiogram and 12-lead electrocardiogram signals. J. Electrocardiol. **42**(6), 622–630 (2009)
26. H. Yang, C. Kan, G. Liu, Y. Chen, Spatiotemporal differentiation of myocardial infarctions. IEEE Trans. Autom. Sci. Eng. **10**(4), 938–947 (2013)
27. R. Bousseljot, D. Kreiseler, A. Schnabel, *Nutzung der EKG-Signaldatenbank CARDIODAT der PTB über das Internet* (1995)
28. A.L. Goldberger, L.A. Amaral, L. Glass, et al., PhysioBank, PhysioToolkit, and PhysioNet: components of a new research resource for complex physiologic signals. Circulation **101**(23), e215–e220 (2000)
29. C. Kan, H. Yang, Dynamic spatiotemporal warping for the detection and location of myocardial infarctions, in *2012 IEEE International Conference on Automation Science and Engineering (CASE)* (IEEE, New York, 2012), pp. 1046–1051

30. H. Yang, C. Kan, A. Krall, D. Finke, Network modeling and Internet of things for smart and connected health systems—a case study for smart heart health monitoring and management. IISE Transactions on Healthcare Systems Engineering **10**(3), 159–171 (2020)
31. C. Kan, F.M. Leonelli, H. Yang, Map reduce for optimizing a large-scale dynamic network-the Internet of Hearts, in *2016 38th Annual International Conference of the IEEE Engineering in Medicine and Biology Society (EMBC)* (IEEE, New York, 2016), pp. 2962–2965
32. C. Kan, H. Yang, Dynamic network monitoring and control of in situ image profiles from ultraprecision machining and biomanufacturing processes. Qual. Reliab. Eng. Int. **33**(8), 2003–2022 (2017)
33. C. Kan, C. Cheng, H. Yang, Heterogeneous recurrence monitoring of dynamic transients in ultraprecision machining processes. J. Manuf. Syst. **41**, 178–187 (2016)
34. Y. Chen, H. Yang, Heterogeneous recurrence t2 charts for monitoring and control of nonlinear dynamic processes, in *2015 IEEE International Conference on Automation Science and Engineering (CASE)* (IEEE, New York, 2015), pp. 1066–1071
35. Y. Chen, G. Liu, H. Yang, Sparse particle filtering for modeling space-time dynamics in distributed sensor networks, in *2014 IEEE International Conference on Automation Science and Engineering (CASE)* (IEEE, New York, 2014), pp. 626–631
36. Y. Chen, H. Yang, Sparse modeling and recursive prediction of space-time dynamics in stochastic sensor networks. IEEE Trans. Autom. Sci. Eng. **13**(1), 215–226 (2015)
37. A. Meyers, H. Yang, Markov chains for fault-tolerance modeling of stochastic networks. IEEE Trans. Autom. Sci. Eng. **19**(3), 2591–2606 (2021)
38. G. Wagner, G. Bub, P. Kohl, F. Pillekamp, Electrocardiography and imaging. J. Electrocardiol. **40**(1S), S66 (2007)
39. B. Huang, W. Kinsner, ECG frame classification using dynamic time warping, in *IEEE CCECE2002: Canadian Conference on Electrical and Computer Engineering. Conference Proceedings (Cat. No. 02CH37373)*, vol. 2 (IEEE, New York, 2002), pp. 1105–1110
40. P.S. Addison, Wavelet transforms and the ECG: A review. Physiol. Meas. **26**(5), R155 (2005)
41. A. Lyon, A. Minchole, J.P. Martinez, P. Laguna, B. Rodriguez, Computational techniques for ECG analysis and interpretation in light of their contribution to medical advances. J. R. Soc. Interface **15**(138), 20170821 (2018)
42. Z. Ebrahimi, M. Loni, M. Daneshtalab, A. Gharehbaghi, A review on deep learning methods for ECG arrhythmia classification. Expert Syst. Appl.: X **7**, 100033 (2020)
43. Y. Xia, H. Zhang, L. Xu, et al., An automatic cardiac arrhythmia classification system with wearable electrocardiogram. IEEE Access **6**, 16529–16538 (2018)
44. B. Zhou, S. Liu, B. Hooi, X. Cheng, J. Ye, BeatGAN: anomalous rhythm detection using adversarially generated time series, in *IJCAI* (2019), pp. 4433–4439
45. P. Wang, B. Hou, S. Shao, R. Yan, ECG arrhythmias detection using auxiliary classifier generative adversarial network and residual network. IEEE Access **7**, 100910–100922 (2019)
46. Z. Wang, S. Stavrakis, B. Yao, Hierarchical deep learning with generative adversarial network for automatic cardiac diagnosis from ECG signals. Comput. Biol. Med. **155**, 106641 (2023)
47. K. Weiss, T.M. Khoshgoftaar, D. Wang, A survey of transfer learning. J. Big Data **3**(1), 1–40 (2016)
48. Z. Wang, B. Yao, Multi-branching temporal convolutional network for sepsis prediction. IEEE J. Biomed. Health Inform. **26**(2), 876–887 (2021)
49. Z. Wang, C. Liu, B. Yao, Multi-branching neural network for myocardial infarction prediction, in *2022 IEEE 18th International Conference on Automation Science and Engineering (CASE)* (IEEE, New York, 2022), pp. 2118–2123
50. S. Chen, Z. Wang, B. Yao, T. Liu, Prediction of diabetic retinopathy using longitudinal electronic health records, in *2022 IEEE 18th International Conference on Automation Science and Engineering (CASE)* (IEEE, New York, 2022), pp. 949–954
51. G.B. Moody, R.G. Mark, The impact of the MIT-BIH arrhythmia database. IEEE Eng. Med. Biol. Mag. **20**(3), 45–50 (2001)
52. T. Li, M. Zhou, ECG classification using wavelet packet entropy and random forests. Entropy **18**(8), 285 (2016)

53. U.R. Acharya, S.L. Oh, Y. Hagiwara, et al., A deep convolutional neural network model to classify heartbeats. Comput. Biol. Med. **89**, 389–396 (2017)
54. S. Mousavi, F. Afghah, F. Khadem, U.R. Acharya, ECG language processing (ELP): a new technique to analyze ECG signals. Comput. Methods Prog. Biomed. **202**, 105959 (2021)
55. M. Kachuee, S. Fazeli, M. Sarrafzadeh, ECG heartbeat classification: a deep transferable representation, in *2018 IEEE International Conference on Healthcare Informatics (ICHI)* (IEEE, New York, 2018), pp. 443–444
56. K. Aras, W. Good, J. Tate, et al., Experimental data and geometric analysis repository-EDGAR. J. Electrocardiol. **48**(6), 975–981 (2015)
57. R.M. Gulrajani, The forward and inverse problems of electrocardiography. IEEE Eng. Med. Biol. Mag. **17**(5), 84–101 (1998)
58. D.H. Brooks, G.F. Ahmad, R.S. MacLeod, G.M. Maratos, Inverse electrocardiography by simultaneous imposition of multiple constraints. IEEE Trans. Biomed. Eng. **46**(1), 3–18 (1999)
59. S. Ghosh, Y. Rudy, Application of l1-norm regularization to epicardial potential solution of the inverse electrocardiography problem. Ann. Biomed. Eng. **37**(5), 902–912 (2009)
60. B. Messnarz, B. Tilg, R. Modre, G. Fischer, F. Hanser, A new spatiotemporal regularization approach for reconstruction of cardiac transmembrane potential patterns. IEEE Trans. Biomed. Eng. **51**(2), 273–281 (2004)
61. L. Wang, H. Zhang, K.C. Wong, H. Liu, P. Shi, Physiological-model-constrained noninvasive reconstruction of volumetric myocardial transmembrane potentials. IEEE Trans. Biomed. Eng. **57**(2), 296–315 (2009)
62. B. Yao, H. Yang, Physics-driven spatiotemporal regularization for high-dimensional predictive modeling: a novel approach to solve the inverse ECG problem. Sci. Rep. **6**(1), 1–13 (2016)
63. B. Yao, H. Yang, Spatiotemporal regularization for inverse ECG modeling. IISE Trans. Healthcare Syst. Eng. **11**(1), 1–13 (2020)
64. J. Xie, B. Yao, Physics-constrained deep learning for robust inverse ECG modeling, in *IEEE Transactions on Automation Science and Engineering* (2022)
65. J. Xie, B. Yao, Physics-constrained deep active learning for spatiotemporal modeling of cardiac electrodynamics. Comput. Biol. Med. **146**, 105586 (2022)
66. R.R. Aliev, A.V. Panfilov, A simple two-variable model of cardiac excitation. Chaos, Solitons & Fractals **7**(3), 293–301 (1996)
67. B. Yao, R. Zhu, H. Yang, Characterizing the location and extent of myocardial infarctions with inverse ECG modeling and spatiotemporal regularization. IEEE J. Biomed. Health Inform. **22**(5), 1445–1455 (2017)
68. B. Yao, R. Zhu, H. Yang, Inverse ECG modeling with spatiotemporal regularization for the characterization of myocardial infarctions, in *2018 IEEE EMBS International Conference on Biomedical and Health Informatics (BHI)* (IEEE, New York, 2018), pp. 120–123
69. F. Dawoud, G.S. Wagner, G. Moody, B.M. Horáček, Using inverse electrocardiography to image myocardial infarction-reflecting on the 2007 PhysioNet/Computers in Cardiology Challenge. J. Electrocardiol. **41**(6), 630–635 (2008)

Chapter 4
Simulation Optimization of Medical Decision Making

Medical decisions (e.g., cardiovascular diagnostics, prognostics, and treatment) are usually made in an uncertain environment, including the time-varying characteristics of diseases, human subject differences, and other noise factors. Optimal medical decision making relies on a better understanding of not only the physics in disease processes but also how it is correlated with the dynamical system behaviors. In the current clinical practice, ad hoc or heuristic strategies are widely used for the medical decisions. For some time, physicians often need to make subjective judgments about the medical decisions, particularly in the presence of high-level uncertainties. This chapter presents a simulation-guided strategy that helps optimize the medical decision making for cardiovascular diseases. Based on simulation models and sensor-based models, the "virtual" heart can be used as a "digital twin" to optimize cardiovascular diagnostics, prognostics, and treatment planning. Simulation-based optimization is strongly promised to optimize medical decisions through the "virtual" heart, as opposed to traditional "trial-and-error" or subjective decisions in the real-world heart.

4.1 Introduction to Simulation Optimization

Simulation optimization is a methodology to find the optimal strategy for a problem when each strategy can be evaluated by running a simulation model instead of physical experiments [1–4]. Advancements in computing power and simulation techniques have expanded simulation optimization as a powerful tool, which has wide applications to improve operational efficiency in healthcare systems, communication networks, environmental systems, and energy systems [1, 5, 6]. Simulation optimization involves searching for specific input settings to a simulation

H. Yang, B. Yao, *Sensing, Modeling and Optimization of Cardiac Systems*,
SpringerBriefs in Service Science, https://doi.org/10.1007/978-3-031-35952-1_4

model such that the objective function (a function of the simulation output) is minimized:

$$\min_{\boldsymbol{\theta} \in \Theta} J(\boldsymbol{\theta}), \tag{4.1}$$

where $\boldsymbol{\theta}$ denotes the decision variable (input parameter setting), Θ is the feasible set that can be discrete, continuous, or a mixture of both, and $J(\boldsymbol{\theta})$ is the objective function. Compared to traditional mathematical programming, solving Eq. (4.1) is more challenging because simulation optimization generally does not assume there is an algebraic description of the simulation. In other words, the simulation may be available as a black box: $J(\boldsymbol{\theta})$ can only be obtained after executing the simulation model given the parameter setting $\boldsymbol{\theta}$. This can be computationally intensive especially for complex and expensive computer simulation, e.g., large-scale cardiac simulation. As such, simulation optimization calls for effective allocation of computing resources among simulation replications under different parameter settings for accurate estimations. A variety of algorithms have been developed to solve the simulation optimization problems under different problem settings.

4.1.1 Rank and Selection

The Ranking & Selection method is commonly used to solve discrete-event simulation optimization problems where the feasible set Θ is not too large [7]. Suppose there are k alternatives and each with the mean performance of μ_x for $x \in \{1, \ldots, k\}$. The goal of Ranking & Selection is to identify the alternative with the best (minimum or maximum) mean value. The Ranking & Selection method can be formulated into two commonly used procedures: fixed-precision and fixed-budget procedures [8]. Fixed-precision procedures aim to achieve a certain probability of selecting correct or good alternatives with a minimum number of simulation replications, and fixed-budget procedures aim to achieve the best estimation within a predefined computational budget.

Two types of statistical models have been developed for Ranking & Selection procedures: frequentist and Bayesian [9]. In frequentist modeling, no prior knowledge on μ_x is required. The mean performance of each alternative and the corresponding variance is estimated by the sample mean and sample variance of the simulation replicates, respectively: $\bar{\mu}_x^n = \frac{1}{n_x}\sum_{i=1}^{n_x} y_i(x)$ and $(\bar{\sigma}_x^n)^2 = \frac{1}{n_x-1}\sum_{i=1}^{n_x}(y_i(x) - \bar{\mu}_x^n)^2$, where n denotes the total number of simulations, $y_i(x)$ is the i-th simulation output given design alternative x, and n_x denotes the simulation replicates for x. Allocating bigger simulation budget to x improves the estimation precision of μ_x at the expense of other alternatives. The Ranking & Selection procedure can then be conducted by evaluating the estimated mean performance of each alternative. On the other hand, in Bayesian modeling, some degree of prior knowledge on the mean performance μ_x is assumed to be known in advance, which is captured by a prior probability distribution. Such prior knowledge may come from historical data, domain knowledge, or preliminary simulations of different alternatives. When additional simulation results become available, the prior knowledge will be refined, which will be represented by a posterior distribution. The Bayesian approach enables to capture the distribution of μ_x with a small set of parameters, significantly reducing the computational burden in Ranking & Selection procedures.

Effective allocation of simulation budget affects the precision of the estimated mean performance, which is critical to the success of Ranking & Selection procedures. A variety of methods have been developed for effective budget allocation to solve the Ranking & Selection problems. For example, Kim and Nelson [10] developed the KN procedure for the frequentist modeling including two stages: in the first stage, each alternative is simulated with equal amount of replicates; the second-stage sampling focuses more on the ones demonstrating high potential to be good candidates and the unpromising ones are screened out. The Bayesian value of information is a commonly used approach for sequential budget allocation designed for the Bayesian modeling procedure, which leverages both the mean estimate and uncertainty quantification to evaluate the potential of a new simulation to improve the quality of the solution [11]. Another commonly used method is the optimal computing budget allocation (OCBA) that can be applied to both frequentist and Bayesian modeling. The OCBA method aims to intelligently determine the simulation replication numbers for all alternatives and optimize the final performance before the simulations are made. Interested readers can refer to [12] for more details.

4.1.2 Response Surface Methodology

The response surface methodology (RSM) [13] is widely used for effective simulation optimization in the continuous problem setting. RSM is a collection of statistical and mathematical techniques aiming to build a functional relationship between the input parameter setting and the output objective. RSM generally consists of two phases [14]. In the first phase, a response surface between the input setting and simulation-based objective is approximated by local first-order polynomials on a small sub-region of the input domain. The coefficients of the polynomials are commonly estimated by least-square methods based on the initial simulation observations of the objective function resulted from design of experiment approaches. Such low-order polynomials are also known as metamodels that run much faster than the simulation models. Given the fitted metamodel, gradient descent will be applied to determine the search direction that is expected to improve the simulation response. Given the new simulation results, the metamodel will be refined. This process iterates multiple times given the computation budget or until a typical convergence criterion is satisfied. In the second phase, the objective function is approximated by higher-order (usually second-order) polynomials, from which the optimal input setting will be estimated by direct differentiation.

A variety of adapted formulations of RSM have been developed to solve different simulation optimization problems. For example, neural-network-based RSMs have been proposed in which the polynomial models in traditional RSMs are replaced by neural network models [15]. Low-cost RSMs have been developed based on central composite and Box–Behnken designs to reduce simulation runs by half compared with standard RSMs while not sacrificing the estimation accuracy [16]. Generalized RSM has been developed to estimate better search direction than the steepest descent direction used in traditional RSM for constrained simulation optimization problems by leveraging mathematical programming techniques [17]. Adaptive RSMs [18] have been built by leveraging a Bayesian metric to select the best response surface from a library of several candidates while considering uncertainty in the data to solve large-scale optimization problems. Dellino et al. proposed to combine Taguchi's view of uncertain world with RSM for robust optimization in simulation [19]. Chang et al. [20] developed stochastic trust region-based RSM to achieve desired convergence properties in simulation optimization.

4.1.3 Stochastic Kriging

Standard RSMs based on regression models provide good functional approximation in local regions, but tend not to perform well as a global response surface over the feasible input space. Global metamodels are of particular interest in simulation optimization especially for real-time decision support. Semiparametric kriging metamodeling has been developed to approximate the global response surface between the input settings and simulation outputs of deterministic models [21, 22]. Specifically, the response surface is modeled as

$$y(\boldsymbol{x}) = \boldsymbol{h}(\boldsymbol{x})^T \boldsymbol{\beta} + w(\boldsymbol{x}), \tag{4.2}$$

Where $y(\boldsymbol{x})$ denotes the simulation response at the design setting $\boldsymbol{x} \in \mathbb{R}^d$, which is noise free in deterministic computer experiments, $\boldsymbol{h}(\boldsymbol{x})^T \boldsymbol{\beta}$ models the mean trend of the response, and $w(\boldsymbol{x})$ is a realization from a zero mean Gaussian random field, which is leveraged to capture the spatial correlation of the responses at different design points. The covariance function is given as

$$Cov[y(\boldsymbol{x}_1), y(\boldsymbol{x}_2)] = Cov[w(\boldsymbol{x}_1), w(\boldsymbol{x}_2)] = \sigma_f^2 \cdot R(\boldsymbol{x}_1, \boldsymbol{x}_2; \boldsymbol{\lambda}) = \Sigma(\boldsymbol{x}_1, \boldsymbol{x}_2), \tag{4.3}$$

Where σ_f^2 denotes the variance of $w(\boldsymbol{x})$ and $R(\cdot, \cdot; \boldsymbol{\lambda})$ is the correlation function with length-scale parameter $\boldsymbol{\lambda}$. The model parameters can be estimated using maximum likelihood estimation based on the initial n_0 simulation observations $\{\boldsymbol{y}, \boldsymbol{X}\}$ with $\boldsymbol{y} = [y_1, y_2, \ldots, y_{n_0}]^T$ and $\boldsymbol{X} = [\boldsymbol{x}_1; \boldsymbol{x}_2; \ldots; \boldsymbol{x}_{n_0}]$. Then, the prediction of simulation response at an arbitrary design point $\boldsymbol{x}$ is provided in Eq. (4.4) without running the computer simulation:

$$\hat{y}(\boldsymbol{x}) = \boldsymbol{h}(\boldsymbol{x})^T \hat{\beta} + \hat{\Sigma}(\boldsymbol{x}, \cdot)\hat{\Sigma}^{-1}(\boldsymbol{y} - \boldsymbol{H}\hat{\boldsymbol{\beta}}), \tag{4.4}$$

where $\boldsymbol{H} = [\boldsymbol{h}(\boldsymbol{x}_1); \boldsymbol{h}(\boldsymbol{x}_2); \ldots; \boldsymbol{h}(\boldsymbol{x}_{n_0})]$. As such, the kriging model can serve as a global response surface for simulation optimization.

Stochastic kriging extends traditional kriging metamodeling to stochastic simulation for response surface approximation [23]. Stochastic kriging is developed to account for both intrinsic (due to the randomness inherent to stochastic simulation) and extrinsic (due to the uncertainty about the predicted response surface) uncertainty in the metamodeling of stochastic simulations [24]. Specifically, the simulation response from at design point $\boldsymbol{x}$ is modeled as

$$y(\boldsymbol{x}) = \boldsymbol{h}(\boldsymbol{x})^T \boldsymbol{\beta} + w(\boldsymbol{x}) + \epsilon, \tag{4.5}$$

where $\epsilon \sim \mathcal{N}(0, \sigma_\epsilon^2)$ is added to model the intrinsic uncertainty. Then, given the maximum likelihood estimation of model parameters, the prediction from stochastic kriging at an unsimulated design point $\boldsymbol{x}$ is

$$\hat{y}(\boldsymbol{x}) = \boldsymbol{h}(\boldsymbol{x})^T \hat{\beta} + \hat{\Sigma}(\boldsymbol{x}, \cdot) \left(\hat{\Sigma} + \hat{\sigma}_\epsilon^2 \cdot \mathcal{I}\right)^{-1} (\boldsymbol{y} - \boldsymbol{H}\hat{\boldsymbol{\beta}}). \tag{4.6}$$

A variety of techniques have been developed to improve the performance of stochastic kriging for stochastic simulation optimization. For example, Chen et al. [25] proposed to improve the response surface prediction by incorporating gradient estimators into stochastic kriging. Che et al. [26] developed an efficient stochastic Kriging by taking the advantage of generalized polynomial chaos. Nested stochastic Kriging [27, 28] has also been developed to improve the accuracy of uncertainty quantification and computation efficiency of simulation optimization.

4.1.4 Simulation Optimization in Healthcare

Simulation optimization approaches have received increasing interests in improving the operational efficiency of service in the healthcare domain [29]. For example, discrete-event simulation optimization has been applied to study resource reallocation problems in healthcare systems for reducing patient waiting time and increasing the level of services [30, 31]. Simulation optimization-based methods including agent-based modeling [32], multi-objective optimization [33], and metamodeling optimization [34] have been proposed to improve the process of emergency department. Mix-integer optimization has been integrated with simulation for optimal inventory management of healthcare supplies [35]. However, most existing works focus on resource allocation, capacity planning, and inventory control. Very little has been done to use simulation optimization to support medical decision making for optimal disease diagnosis and treatment.

4.2 Sequential Medical Decision Making

Rapid advances in medical sciences and technology innovations have significantly improved the information availability in the healthcare diagnostics and treatment environment. Such advancements, in turn, complicate the medical decision making because of the exploding information streams and a growing number of alternative options [36, 37]. Additionally, medical decision making often involves a sequential process [38], where the decisions are often made sequentially in a stochastic environment and are dynamically interdependent over time. For example, decision making of medical intervention for chronic disease is a dynamic sequential process over long periods that can be as long as the patient's life [39]. Thus, there exist uncertainties in the evolution of the disease dynamics and in the treatment effects on the patient's health condition over the long term. Various factors impact the effects of medical treatments including genetic, dietary, and environmental differences. Undergoing the intervention at a different time or in a different environment will generally result in variations in treatment effects. Optimal medical decision making calls for incorporating relevant information such as the patient's health status and alternative clinical options into a sequential optimization process.

4.2.1 Model-Based Sequential Decision Making

Sequential decision making problems can be formulated as a model-free or model-based framework. In the model-based sequential decision framework, a prior model is selected to describe the transition dynamics among the states based on clinical domain knowledge and statistical learning of historical data. The Markov decision process (MDP) [40–42] is a commonly used framework and has been extensively studied for optimal sequential management of chronic diseases such as breast cancer and diabetes [43–47]. The MDP model is generally characterized by a 4-tuple $(\mathcal{S}, \mathcal{A}, \mathcal{P}, r)$, where $\mathcal{S} = \{s_1, \ldots, s_T\}$ denotes the state space, $\mathcal{A}$ is the action space composed by the alternative options for medical intervention or treatment, $\mathcal{P}$ is a transition matrix describing the system dynamics, and $r : \mathcal{S} \times \mathcal{A} \rightarrow \mathbb{R}$ defines the reward function for each state–action pair. The optimal policy can be further derived by estimating the value function of each state, which is defined as the expected cumulative reward over the planning horizon T. Note that dynamic programming or value iteration algorithm is widely used to estimate the value function (especially for finite-horizon MDP), because the transition dynamics is assumed to be known and characterized by $\mathcal{P}$ in model-based MDP:

$$v(s_t) = \max_{a_t \in \mathcal{A}} \{r(s_t, a_t) + \sum_{s_{t+1} \in \mathcal{S}} \mathcal{P}(s_{t+1}|s_t, a_t) v(s_{t+1})\}, \quad t = 1, \ldots, T-1 \tag{4.7}$$

$$v(s_T) = r(s_t). \tag{4.8}$$

The fundamental challenge in model-based MDP methods lies in estimating the transition probabilities, which is required for every possible state–action pair. The transition dynamics can be estimated from clinical domain knowledge, simulation modeling, or longitudinal healthcare datasets. For example, Abdollahian and Das [46] utilized the survival analysis data [48] derived from the SEER database to estimate the incidence probabilities of various stages of cancers for BRCA mutation carriers and further solved the MDP for optimal intervention strategies. Ayer et al. [49] estimated the transition probabilities using a validated micro-simulation model and solved the partially observable MDP for personalized screening policy of mammography for patients with breast cancer. Oh et al. [47] estimated the transition probabilities by counting the number of occurrences between two states under different medications from a population-based cohort and further solved the MDP for optimal treatment recommendations for diabetes patients. However, accurately estimating the transition dynamics will be infeasible if dimensionality of the state or action space is large, which is also known as the curse of dimensionality [50].

4.2.2 Model-Free Sequential Decision Making

Reinforcement learning is a commonly used model-free method that requires no specific assumptions of the underlying stochastic model or transition probabilities. It has a wide application to optimize sequential medical decision making problems such as dynamic treatment regimes for chronic diseases and automated medical diagnosis [51]. In reinforcement learning, instead of explicitly learning a model of the system, an artificial agent learns to create an optimal strategy through direct interactions with the system: trying different sequences of actions, observing the outcomes of those actions, and choosing the one that yields the most desirable outcome. The random process resulted from the action sequence over T time periods is characterized by the following trajectory:

$$\{s_0, (s_1, a_1, r_{s_1,a_1}), \ldots, (s_{T-1}, a_{T-1}, r_{s_{T-1},a_{T-1}}), (s_T, r_{s_T})\}, \tag{4.9}$$

where s_t denotes the system state at period t, a_t is the action (e.g., medical intervention or treatment) selected at t, and r_{s_t,a_t} denotes the reward incurred by the state–action pair (s_t, a_t). The observed trajectories under an arbitrary policy π will be used to estimate the value function (i.e., cumulative reward over the planning horizon) of each state:

$$v^{\pi}(s_t) = E_{\pi}\left[\sum_{\tau=1}^{T-1} r^{\pi}_{s_\tau, a_\tau} + r^{\pi}_{s_T} | s_t\right]. \tag{4.10}$$

Reinforcement learning aims to estimate the value function, which can be achieved by the temporal difference learning or Q-learning algorithm, and further derive the optimal policy that maximizes the estimated value function.

In many real-world problems with large state and action spaces, model-free reinforcement learning for the optimal policy from scratch is intractable because extensive computational resource is needed to sufficiently evaluate each policy. To cope with this challenge, a variety of Bayesian reinforcement learning approaches [52, 53] have been developed for efficiently searching for the optimal policy. In Bayesian reinforcement learning, a prior distribution derived from the domain knowledge or historical data is assigned over the Q-function or value function to reduce the number of interactions with the system needed by the artificial agent for searching for the optimal policy. Chang et al. [54] have further developed an adaptive sampling algorithm inspired from the multi-armed bandit problem to balance the tradeoff between the exploration and exploitation for improving the convergence rate in reinforcement learning. Note that the model-free reinforcement learning problem can also be represented with a decision tree, where each node indicates the current state and the edges coming out of the node represent possible actions (decisions) to take at the state. The combination of state and action will generate a reward and determines the next state (node of the tree). Each path of the tree expands until it reaches a leaf node (e.g., end of the planning horizon). Monte Carlo tree search-based algorithms have been developed for the efficient exploration and evaluation of the paths on the tree to solve large sequential decision making problems [55, 56].

4.3 Optimal Cardiac Surgical Planning

Many patients have to face surgical interventions to fight the battle against heart disease. In the USA, the number of cardiac procedures performed is around 51.4 million annually. Surgical successes are of paramount importance for the patients' health and their family well-being, thereby broadly impacting the society. For example, atrial fibrillation (AF) is the most common arrhythmia in elder population, affecting around 2.7 million Americans [57]. Catheter ablation is an established surgery procedure for AF, which creates incision lines to block faulty electrical pathways [58]. However, there are large variations in surgical outcomes due to the disease complexity and patient-specific variations, as well as the dynamic cardiac responses under different treatments [59]. Cardiac surgery generally involves sequential procedures, and the response of the heart evolves over time when clinicians take actions during the surgical operation. A predefined surgical procedure is static and cannot adapt to cardiac changes in intermediate surgical steps. Optimal surgery planning calls for formulating it as a sequential decision making problem.

Cardiac simulation provides an unprecedented opportunity to evaluate alternative surgery procedures, which can be integrated with sequential optimization methods to search for the optimal surgery plan. However, traditional sequential decision making framework such as model-based MDP or model-free reinforcement learning is not applicable in the context of cardiac surgery planning due to the following reasons: (1) Functional simulation output—traditional sequential optimization algorithm is commonly applied to simulation models with scalar—or vector-valued outputs. However, the cardiac simulator generates functional outputs, e.g., electrical signals as a function of time and location in the heart. (2) Complex 3D geometry—the state and action spaces lie on the 3D heart geometry. For example, AF ablation involves creating incision lines on the heart surface by radiofrequency catheters to destroy faulty electrical pathways. The geometric or topology information needs to be incorporated to evaluate the effectiveness of alternative surgery paths. (3) High-dimensional action space—sequential planning of cardiac surgery often involves a large action space, which can be any arbitrary surgical patterns (e.g., ablation sequences or incision lines in AF ablation) over the heart geometry. Novel sequential simulation optimization methods are needed to cope with the aforementioned challenges for optimal cardiac surgery planning.

4.3.1 Sequential Decision Making Formulation of Cardiac Surgery Problems

We formulate the sequential optimization problem for cardiac surgery planning with the following 4-tuple $(\mathcal{S}, \mathcal{A}, \mathcal{T}, \mathcal{R})$:

- **State**: $s_t \in \mathcal{S}$, where $\mathcal{S}$ denotes the set of feasible states that can potentially take the surgery action to cure the heart disease; s_t is a size-n 0–1 vector whose element is 1 if the corresponding spatial node has been treated at time t, and 0, otherwise. Note that n denotes the total number of nodes in the heart geometry. Then, the state space can be defined as $\mathcal{S} = \{s \in \{0, 1\}^n \,|\, \|s\|_1 \leq H\}$, where H is the maximum allowed number of surgery steps.
- **Action**: $a_t \in \mathcal{A} \subseteq \{0, 1\}^n$ is a unit vector with entry 1 corresponding to the spatial node taking the surgery at time t and 0, elsewhere. The feasible action space given state s_t, denoted by $\mathcal{A}(s_t)$, is the remaining points that have not yet been treated at time t.
- **State transition**: The state transition function, $\mathcal{T} : \mathcal{S} \times \mathcal{A}$, is defined as $s_{t+1} = \mathcal{T}(s_t, a_t) = s_t + a_t$; s_{t+1} is updated by adding the newly treated points at t to s_t.

- **Reward**: The reward function, $R : \mathcal{S} \times \mathcal{A} \to \{-1, 0\}$, is denoted as $R(s_t, a_t)$, where $R(s_t, a_t) = 0$ if the heart disease is cured under action a_t at state s_t, and otherwise, $R(s_t, a_t) = -1$.

The optimal policy is then defined as the one that maximizes the cumulative reward:

$$\pi^* = \underset{\{\pi:\pi(s_t)=a_t\in\mathcal{A}(s_t)\}}{\arg\max} \sum_{t=0}^{H} R(s_t, a_t).$$

Note that $-\sum_{t=0}^{H} R(S_t, a_t)$ is the number of treated points until either the heart disease is cured or H is reached. Hence, the optimal policy is the one that cures the disease while causing the least damage (i.e., treated nodes) to the heart.

MCTS is a widely used algorithm to solve the sequential optimization problem. Each round of MCTS consists of four phases: (1) **Selection**—Starting from the beginning of the surgery, we select successive points to treat until reaching a state where no further treatment is simulated, which is referred to as a leaf node; (2) **Expansion**—one action is selected from the leaf node; (3) **Rollout**—Starting from the leaf node, subsequent treatments are simulated until either the disease is cured or H is reached; (4) **Backpropagation**—the value functions of nodes in the tree are updated to reflect the newly simulated result. These four phases are repeated until the computation budget runs out.

During the MCTS exploration, the agents need to balance the exploitation of the optimal action given the current information with the exploration of other actions that currently appear suboptimal but may turn out to be superior in the long run. The UCT (the upper confidence bounds applied to tree) algorithm [54, 60] is a common action-selection strategy in MCTS to balance such exploitation and exploration. Specifically, each node of the tree is treated as a multi-armed bandit, and the action is selected according to

$$a^*(s_t) = \underset{a_t}{\arg\max} \left\{ Q(s_t, a_t) + c\sqrt{\log N(s_t)/N(s_t, a_t)} \right\},$$

where c is a constant (usually equals $\sqrt{2}$) to balance the exploitation and exploration, $N(s_t, a_t)$ is the number of visits of taking action a_t in state s_t, and $N(s_t) = \sum_{a_t} N(s_t, a_t)$. Finally, $Q(s_t, a_t)$ is the cumulative reward from subsequent actions starting at s_t, if action a_t is taken:

$$Q(s_t, a_t) = R(s_t, a_t) + V(\mathcal{T}(s_t, a_t)),$$

where $V(s_t) = \max_{a_t\in\mathcal{A}(s_t)} Q(s_t, a_t)$ denotes the value function of state s_t. Note that MCTS for cardiac surgery is challenged by not only the large state and action spaces but also the expensive cardiac simulation. To address such challenges, our prior work has developed the Bayesian-enhanced MCTS algorithm to effectively tackle this problem.

4.3.2 Bayesian Learning-Enhanced Tree Search for Optimal Cardiac Surgical Planning

Bayesian methods have been developed to infer the value function of a state–action pair that has not been simulated yet for sequential decision making problems with large action spaces [61]. In general, Bayesian function approximation starts with non-informative prior knowledge about the function and updates the posterior conditioned on the inputs and corresponding outputs. Note that the correlation kernel is critical to the success of Bayesian methods: if the two inputs are similar, then the corresponding responses are highly correlated. Similarity of the inputs is typically measured by a distance metric in the input space. In our context, we model the reward of a surgery pattern using Bayesian function approximation, which requires a measure of distance between two ablation patterns. Specifically, we use the following kernel:

$$K(s_2, s_2) = \exp\left(-\frac{d_{\mathcal{H}}^2(s_1, s_2)}{2\lambda_1^2} - \frac{(\|s_1\|_1 - |s_2\|_1)^2}{2\lambda_2^2}\right), \tag{4.11}$$

where $d_{\mathcal{H}}(s_1, s_2)$ is the Hausdorff distance to measure the similarity between two surgery patterns s_1 and s_2, and second component inside the exponential measures the contribution of the length of the surgery pattern to the kernel. However, a kernel purely based on Hausdorff distance is not positive definite. We use the representative set method [62] in our prior work to build a valid kernel for Bayesian learning. Using the valid kernel, we train a Bayesian model to predict the reward of a state–action pair that is not yet simulated based on already simulated states for efficient MCTS.

In the Bayesian-enhanced MCTS approach, we denote the predictive distribution of reward for state–action pair $(\boldsymbol{s}, \boldsymbol{a})$ as $r(\boldsymbol{s}, \boldsymbol{a})$, which can be computed from the Bayesian regression model (e.g., GP model) even if $(\boldsymbol{s}, \boldsymbol{a})$ has not been simulated yet. As such, we can estimate the Q-value, $Q(\boldsymbol{s}, \boldsymbol{a})$, using the following recursion:

$$\hat{Q}(\boldsymbol{s}, \boldsymbol{a}) = E[r(\boldsymbol{s}, \boldsymbol{a})] + \hat{V}(\mathcal{T}(\boldsymbol{s}, \boldsymbol{a})), \qquad \hat{V}(\boldsymbol{s}) = \max_{\boldsymbol{a} \in \mathcal{A}(\boldsymbol{s})} \hat{Q}(\boldsymbol{s}, \boldsymbol{a}). \tag{4.12}$$

Given $\hat{Q}(\boldsymbol{s}, \boldsymbol{a})$, we propose the following criterion to select action $\boldsymbol{a}^*$ at state $\boldsymbol{s}$ in the selection phase of MCTS:

$$a^*(s) = \arg\max_{a \in \mathcal{A}(s)} \left\{ \hat{Q}(s, a) + c_{select} \sqrt{\frac{\ln N_s}{n(\boldsymbol{s}, \boldsymbol{a}) + 1}} \right\}, \tag{4.13}$$

where N_s is the number of times state $\boldsymbol{s}$ has been visited and $n(\boldsymbol{s}, \boldsymbol{a})$ is the number of times $\boldsymbol{a}$ is selected at $\boldsymbol{s}$. As such, we derive $\hat{Q}(\boldsymbol{s}, \boldsymbol{a})$ from our Bayesian prediction model instead of estimating it by a sample average as in traditional MCTS algorithms to improve the computational efficiency.

We have implemented the Bayesian-enhanced MCTS (i.e., GP-MCTS) for optimal ablation planning of AF in a simulation study. The objective is to design optimal ablation path for restoring normal sinus rhythm (NSR) from AF. The optimal ablation path is successfully identified to destroy AF signals and outperforms the clinical ablation procedure, i.e., PVI, as shown in Fig. 4.1. Figure 4.2 further compares the NSR and AF signals ablated by the Bayesian-enhanced MCTS policy and PVI given two AF triggers. The ablated AF signals from our method closely match NSR for both AF triggers. However, the PVI procedure only successfully blocks trigger 1 to restore NSR (Fig. 4.2a) but fails to do so (Fig. 4.2b) when the trigger is far away from the pulmonary veins (i.e., trigger 2).

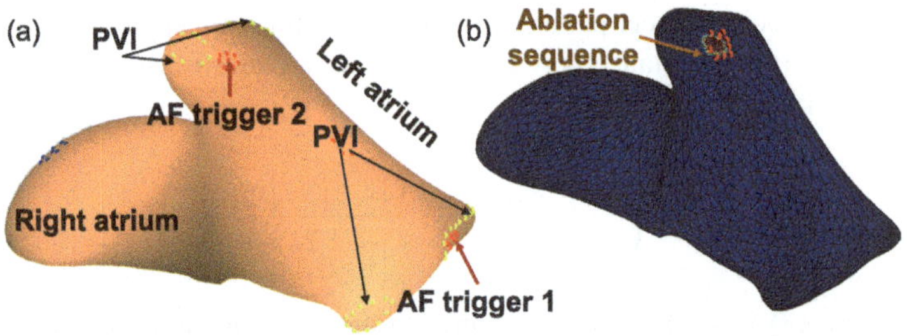

Fig. 4.1 (**a**) Clinical ablation (i.e., PVI); (**b**) Ablation path identified by GP-MCTS with trigger 2

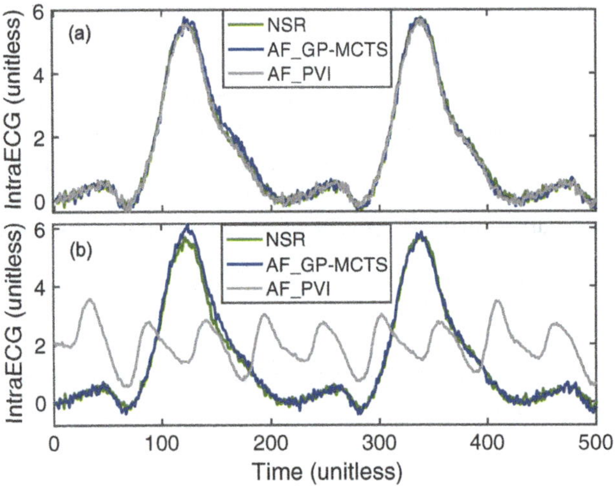

Fig. 4.2 Comparison of NSR and ablated AF signals from GP-MCTS and PVI with (**a**) triggers 1 and (**b**) 2

4.4 Conclusions

This chapter highlights the role of simulation optimization and sequential analytics in medical decision making. We first introduce the general framework of simulation optimization and its various applications in the healthcare domain. Second, we present the commonly used methods including model-based MDP and model-free reinforcement learning for sequential medical decision making. Finally, we present a novel sequential simulation optimization framework of Bayesian learning-enhanced MCTS for optimal cardiac surgery planning. We hope to emphasize the importance of integrating advanced data analytics, cardiac computation modeling, and simulation optimization to the acceleration and realization of precision cardiology.

References

1. M.C. Fu, F.W. Glover, J. April, Simulation optimization: a review, new developments, and applications, in *Proceedings of the Winter Simulation Conference, 2005* (IEEE, New York, 2005), 13-pp.
2. E. Tekin, I. Sabuncuoglu, Simulation optimization: a comprehensive review on theory and applications. IIE Trans. **36**(11), 1067–1081 (2004)
3. S. Amaran, N.V. Sahinidis, B. Sharda, S.J. Bury, Simulation optimization: a review of algorithms and applications. Ann. Oper. Res. **240**(1), 351–380 (2016)
4. P.L. Salemi, E. Song, B.L. Nelson, J. Staum, Gaussian Markov random fields for discrete optimization via simulation: framework and algorithms. Oper. Res. **67**(1), 250–266 (2019)
5. Y. Carson, A. Maria, Simulation optimization: methods and applications, in *Proceedings of the 29th Conference on Winter Simulation* (1997), pp. 118–126
6. M. Hoffman, E. Song, M.P. Brundage, S. Kumara, Online improvement of condition-based maintenance policy via Monte Carlo tree search. IEEE Trans. Autom. Sci. Eng. **19**(3), 2540–2551 (2021)
7. S.-H. Kim, B.L. Nelson, Recent advances in ranking and selection, in *2007 Winter Simulation Conference* (IEEE, New York, 2007), pp. 162–172
8. L.J. Hong, W. Fan, J. Luo, Review on ranking and selection: a new perspective. Frontiers of Engineering Management **8**(3), 321–343 (2021)
9. C.-H. Chen, S.E. Chick, L.H. Lee, N.A. Pujowidianto, Ranking and selection: efficient simulation budget allocation, in *Handbook of Simulation Optimization* (2015), pp. 45–80
10. S.-H. Kim, B.L. Nelson, Selecting the best system, in *Handbooks in Operations Research and Management Science*, vol. 13 (2006), pp. 501–534
11. S.E. Chick, Subjective probability and Bayesian methodology, in *Handbooks in Operations Research and Management Science*, vol. 13 (2006), pp. 225–257
12. L.H. Lee, C.-h. Chen, E.P. Chew, J. Li, N.A. Pujowidianto, S. Zhang, A review of optimal computing budget allocation algorithms for simulation optimization problem. International Journal of Operations Research **7**(2), 19–31 (2010)
13. A.I. Khuri, S. Mukhopadhyay, Response surface methodology. Wiley Interdiscip. Rev. Comput. Stat. **2**(2), 128–149 (2010)
14. H.G. Neddermeijer, G.J. Van Oortmarssen, N. Piersma, R. Dekker, A framework for response surface methodology for simulation optimization, in *2000 Winter Simulation Conference Proceedings (Cat. No. 00CH37165)*, vol. 1 (IEEE, New York, 2000), pp. 129–136
15. M.F. Anjum, I. Tasadduq, K. Al-Sultan, Response surface methodology: a neural network approach. Eur. J. Oper. Res. **101**(1), 65–73 (1997)

16. T.T. Allen, L. Yu, Low-cost response surface methods from simulation optimization. Qual. Reliab. Eng. Int. **18**(1), 5–17 (2002)
17. J.P. Kleijnen, Response surface methodology for constrained simulation optimization: an overview. Simul. Model. Pract. Theory **16**(1), 50–64 (2008)
18. L. Shi, R.-J. Yang, P. Zhu, An adaptive response surface method for crashworthiness optimization. Eng. Optim. **45**(11), 1365–1377 (2013)
19. G. Dellino, J.P. Kleijnen, C. Meloni, Robust optimization in simulation: Taguchi and response surface methodology. Int. J. Prod. Econ. **125**(1), 52–59 (2010)
20. K.-H. Chang, L.J. Hong, H. Wan, Stochastic trust-region response-surface method (strong)—a new response-surface framework for simulation optimization. INFORMS J. Comput. **25**(2), 230–243 (2013)
21. H. Jalali, I. Van Nieuwenhuyse, V. Picheny, Comparison of Kriging-based algorithms for simulation optimization with heterogeneous noise. Eur. J. Oper. Res. **261**(1), 279–301 (2017)
22. S. Rojas-Gonzalez, I. Van Nieuwenhuyse, A survey on Kriging-based infill algorithms for multiobjective simulation optimization. Comput. Oper. Res. **116**, 104869 (2020)
23. B. Ankenman, B.L. Nelson, J. Staum, Stochastic Kriging for simulation metamodeling, in *2008 Winter Simulation Conference* (IEEE, New York, 2008), pp. 362–370
24. J. Staum, Better simulation metamodeling: the why, what, and how of stochastic Kriging, in *Proceedings of the 2009 Winter Simulation Conference (WSC)* (IEEE, New York, 2009), pp. 119–133
25. X. Chen, B.E. Ankenman, B.L. Nelson, Enhancing stochastic Kriging metamodels with gradient estimators. Oper. Res. **61**(2), 512–528 (2013)
26. Y. Che, Z. Guo, C. Cheng, Generalized polynomial chaos-informed efficient stochastic Kriging. J. Comput. Phys. **445**, 110598 (2021)
27. M. Liu, J. Staum, Stochastic Kriging for efficient nested simulation of expected shortfall. J. Risk **12**(3), 3 (2010)
28. P. Hao, S. Feng, H. Liu, Y. Wang, B. Wang, B. Wang, A novel nested stochastic Kriging model for response noise quantification and reliability analysis. Comput. Methods Appl. Mech. Eng. **384**, 113941 (2021)
29. L. Wang, E. Demeulemeester, Simulation optimization in healthcare resource planning: a literature review. IISE Transactions, no. just-accepted, 1–78 (2022)
30. T.-H. Hejazi, State-dependent resource reallocation plan for health care systems: a simulation optimization approach. Comput. Ind. Eng. **159**, 107502 (2021)
31. M. Ordu, E. Demir, C. Tofallis, M.M. Gunal, A novel healthcare resource allocation decision support tool: a forecasting-simulation-optimization approach. J. Oper. Res. Soc. **72**(3), 485–500 (2021)
32. E. Cabrera, M. Taboada, M.L. Iglesias, F. Epelde, E. Luque, Simulation optimization for healthcare emergency departments. Procedia Computer Science **9**, 1464–1473 (2012)
33. A.G. Uriarte, E.R. Zúñiga, M.U. Moris, A.H. Ng, How can decision makers be supported in the improvement of an emergency department? a simulation, optimization and data mining approach. Oper. Res. Health Care **15**, 102–122 (2017)
34. A. Bahari, F. Asadi, A simulation optimization approach for resource allocation in an emergency department healthcare unit. Glob. Heart **15**(1), 1–6 (2020)
35. M. Buschiazzo, J. Mula, F. Campuzano-Bolarin, Simulation optimization for the inventory management of healthcare supplies. Int. J. Simul. Model. **19**(2), 255–266 (2020)
36. B.T. Denton, Frontiers of medical decision-making in the modern age of data analytics. IISE Trans. **55**(1), 94–105 (2023)
37. B.T. Denton, O. Alagoz, A. Holder, E.K. Lee, Medical decision making: open research challenges. IIE Transactions on Healthcare Systems Engineering **1**(3), 161–167 (2011)
38. S. Tunc, O. Alagoz, E. Burnside, Opportunities for operations research in medical decision making. IEEE Intell. Syst. **29**(3), 59 (2014)
39. L.N. Steimle, B.T. Denton, Markov decision processes for screening and treatment of chronic diseases, in *Markov Decision Processes in Practice* (Springer, Berlin, 2017), pp. 189–222

40. B. Yao, Y. Chen, H. Yang, Constrained Markov decision process modeling for optimal sensing of cardiac events in mobile health. IEEE Trans. Autom. Sci. Eng. **19**(2), 1017–1029 (2021)
41. B. Yao, H. Yang, Constrained Markov decision process modeling for sequential optimization of additive manufacturing build quality. IEEE Access **6**, 54786–54794 (2018)
42. B. Yao, F. Imani, H. Yang, Markov decision process for image-guided additive manufacturing. IEEE Rob. Autom. Lett. **3**(4), 2792–2798 (2018)
43. A.J. Schaefer, M.D. Bailey, S.M. Shechter, M.S. Roberts, Modeling medical treatment using Markov decision processes, in *Operations Research and Health Care* (Springer, Berlin, 2005), pp. 593–612
44. J. Chhatwal, O. Alagoz, E.S. Burnside, Optimal breast biopsy decision-making based on mammographic features and demographic factors. Oper. Res. **58**(6), 1577–1591 (2010)
45. F. Imani, Z. Qiu, H. Yang, Markov decision process modeling for multi-stage optimization of intervention and treatment strategies in breast cancer, in *2020 42nd Annual International Conference of the IEEE Engineering in Medicine and Biology Society (EMBC)* (IEEE, New York, 2020), pp. 5394–5397
46. M. Abdollahian, T.K. Das, A MDP model for breast and ovarian cancer intervention strategies for BRCA1/2 mutation carriers. IEEE J. Biomed. Health Inform. **19**(2), 720–727 (2014)
47. S.-H. Oh, S.J. Lee, J. Noh, J. Mo, Optimal treatment recommendations for diabetes patients using the Markov decision process along with the South Korean electronic health records. Sci. Rep. **11**(1), 1–10 (2021)
48. A.W. Kurian, B.M. Sigal, S.K. Plevritis, Survival analysis of cancer risk reduction strategies for BRCA1/2 mutation carriers. J. Clin. Oncol. **28**(2), 222 (2010)
49. T. Ayer, O. Alagoz, N.K. Stout, OR forum—a POMDP approach to personalize mammography screening decisions. Oper. Res. **60**(5), 1019–1034 (2012)
50. W.B. Powell, *Approximate Dynamic Programming: Solving the Curses of Dimensionality*, vol. 703 (Wiley, New York, 2007)
51. C. Yu, J. Liu, S. Nemati, G. Yin, Reinforcement learning in healthcare: A survey. ACM Comput. Surv. (CSUR) **55**(1), 1–36 (2021)
52. M. Ghavamzadeh, S. Mannor, J. Pineau, A. Tamar, et al., Bayesian reinforcement learning: A survey. Found. Trends® Mach. Learn. **8**(5–6), 359–483 (2015)
53. N. Vlassis, M. Ghavamzadeh, S. Mannor, P. Poupart, Bayesian reinforcement learning, in *Reinforcement Learning* (2012), pp. 359–386
54. H.S. Chang, M.C. Fu, J. Hu, S.I. Marcus, An adaptive sampling algorithm for solving Markov decision processes. Oper. Res. **53**(1), 126–139 (2005)
55. T. Vodopivec, S. Samothrakis, B. Ster, On Monte Carlo tree search and reinforcement learning. J. Artif. Intell. Res. **60**, 881–936 (2017)
56. A. Guez, D. Silver, P. Dayan, Scalable and efficient Bayes-adaptive reinforcement learning based on Monte-Carlo tree search. J. Artif. Intell. Res. **48**, 841–883 (2013)
57. D.J. Copley, K.M. Hill, Atrial fibrillation: a review of treatments and current guidelines. AACN Adv. Crit. Care **27**(1), 120–128 (2016)
58. O. Wazni, B. Wilkoff, W. Saliba, Catheter ablation for atrial fibrillation. N. Engl. J. Med. **365**(24), 2296–2304 (2011)
59. R.A. Winkle, R.H. Mead, G. Engel, R.A. Patrawala, Long-term results of atrial fibrillation ablation: the importance of all initial ablation failures undergoing a repeat ablation. Am. Heart J. **162**(1), 193–200 (2011)
60. C.B. Browne, E. Powley, D. Whitehouse, et al., A survey of Monte Carlo tree search methods. IEEE Trans. Comput. Intell. AI Games **4**(1), 1–43 (2012)
61. G. Tesauro, V. Rajan, R. Segal, Bayesian inference in Monte Carlo tree search. arXiv preprint arXiv:1203.3519 (2012)
62. L. Wu, I.E.-H. Yen, F. Xu, P. Ravikumar, M. Witbrock, D2KE: from distance to kernel and embedding. arXiv preprint arXiv:1802.04956 (2018)

Chapter 5
Outlook and Future Research

Modern healthcare systems are investing heavily in advanced sensing and imaging technology to increase the ability of capturing extensive medical data from patients, which is conducive to improving information visibility and coping with disease complexity. On the other hand, simulation modeling of the heart provides a fundamental infrastructure for the visualization and quantitative understanding of spatiotemporal cardiac dynamic processes. Precision cardiology calls upon the development of advanced computation models that are able to effectively promote and leverage the synergy between physics-based cardiac simulation and data-driven machine learning models to exploit the medical data from different modalities, further constructing a digital twin of the heart. Such computational models will enable the integrative analysis of a patient's medical data, low-cost and low-risk quantitative evaluation of alternative therapeutic designs, and generating diagnostic decisions that are free from subjective assessment.

Potential future research directions to further promote the wide adoption of cardiac computational models in clinical practice are listed as follows:

Computational Efficiency in Cardiac Simulation Applications of cardiac simulation in real clinical practice require personalizing the models to patient-specific data, which need conducting a large number of simulation runs. The physics-based computational models are generally solved iteratively using finite element methods (FEMs), which subdivide the large and complex system into smaller and simpler parts for numerical solutions. Specifically, the propagation of cardiac electrodynamics is derived by numerically solving the partial differential equations that govern the cardiac dynamic process through the discretization of the spatial and temporal domains. Note that FEM requires sophisticated mesh-generation techniques prone to convergence issues and requires significant computing resources. For example, anatomically detailed cardiac simulation could take hours or even days for a single-cycle simulation. The considerable computational expense significantly limits the clinical application of FEM-based cardiac simulation in real-time prediction and analysis of the pharmaceutical effects and therapeutic outcomes. Significant

H. Yang, B. Yao, *Sensing, Modeling and Optimization of Cardiac Systems*,
SpringerBriefs in Service Science, https://doi.org/10.1007/978-3-031-35952-1_5

research efforts need to be invested in designing advanced parallel computing techniques and efficient computational infrastructure to increase the scalability of cardiac simulation.

Multi-modal Sensor Data Fusion Physiological sensing and imaging data describing the electrical activities of the heart are often collected from multiple sources with a high level of heterogeneity. For example, electrodes are placed both on the body surface and in the heart chamber to capture dynamic cardiac responses during surgery procedures, generating intracardiac electrograms and surface ECG signals. Intracardiac echocardiography is another promising ultrasound-based technique, generating imaging data of the detailed intracardiac structures. Such multi-source signals need to be modeled together because the single-source data may not be informative enough for decision making. Additionally, human heart is a nonlinear dynamical system that works in real time to support vital living organs. Real-time physiological sensing brings dynamic data streams associated with the cardiac function, which needs to be assimilated into cardiac computational modeling for sequential surgical plans and treatment design. Research efforts are urgently needed to effectively handle the complex data structure (e.g., dynamic signals distributed on complex 3D geometries), intelligently fuse sensor data from multiple sources, and efficiently assimilate dynamic signals in real time, further bringing cardiac computational modeling closer to clinical practice.

Design of Computer Experiments in Cardiac Simulation Simulation-based optimal surgical planning or therapeutic design requires the evaluation of alternative options on the virtual heart. Due to the considerable computation overhead, it is infeasible to exhaustively evaluate every possible design through cardiac simulation. As such, effective design of computer experiments is needed for simulation-based optimal treatment plan for heart disease. Although classical design of experiments (DOE) methods has been extensively studied in the literature, their straightforward application to cardiac simulation is inappropriate due to the following fundamental challenges: (1) Functional cardiac response—cardiac response under different treatment plans is a spatiotemporal variable. The metric used to evaluate the effectiveness of alternative plans needs to be carefully defined to account for the spatiotemporal data structure. (2) Complex design space—the surgical procedure to treat heart disease is often applied on the 3D heart chamber. For example, AF ablation involves creating incision lines on the heart surface by radiofrequency catheters to destroy faulty electrical pathways. The 3D geometric information of the heart needs to be incorporated to evaluate the effectiveness of alternative treatment plans. As such, significant research efforts need to be invested in developing advanced DOE methodologies that are able to not only account for the dynamic cardiac response and complex heart geometry but also balance the exploration and exploitation dilemma to fully utilize cardiac simulation for optimal treatment development.

The manufacturer's authorised representative in the EU is Springer Nature Customer Service Centre GmbH, Europaplatz 3, 69115 Heidelberg, Germany. If you have any concerns regarding our products, please contact ProductSafety@springernature.com

Printed and bound by CPI Group (UK) Ltd, Croydon, CR0 4YY
01/10/2026
02225880-0006